Bomb Safety and Security

The Manager's Guide

Donald S. Williams CPP

National Library of Australia Cataloguing-in-Publication entry

Author: Williams, Donald S. 1957-

Title: Bomb Safety and Security – The Manager's Guide

ISBN: 978-0-9757873-7-3 Print
ISBN: 978-0-9804337-4-6 eBook

Series: Security and resilience series.

Notes: Includes bibliographical references.

Subjects: Improvised explosive devices.

           Blast effect.

           Blast injuries.

           Buildings--Blast effects.

Dewey Number: 363.325

The Australian Security Research Centre is a think tank working in the areas of national security and emergency management. The Centre undertakes independent, evidence-based analysis of both traditional and evolving security, and emergency management issues. Collaborative Publications (C-Pubs) is an Australian royalty publisher that publishes quality non-fiction, niche books.

www.asrc.com.au
www.c-pubs.com.au

ASRC Publication Series

The Australian Security Research Centre is a strategic, public policy centre that originates and develops applicable research into issues related to security in the broadest sense. To promulgate the results of the research and the individual expertise of its associates the ASRC publishes authoritative references that are designed to be applied by those with an interest and responsibility for security and related managerial disciplines.

Author's Foreword

This book had its origins in the mid-1980s as I watched the head of a bomb squad deliver a lecture on how to build a bomb to a room full of facility managers who desperately wanted to know what to do when they found one. It raised the question "What do managers need?" This book is the culmination of over thirty years of thought seeking to answer that question. The intent is to help those responsible for protecting people, property and productivity.

Bomb incidents, be they threats, unattended items, bombs or post-blast, are real if unfortunate parts of the security spectrum. It is the responsibility of managers to be able to safely and effectively identify and respond to such incidents. This book is designed for government and corporate managers whose primary consideration is how to protect life and the organization while minimizing unnecessary disruption.

There have been many who have influenced my thinking over the years, some through providing a new thought or concept, some through encouragement and support and some by proposing concepts and procedures that led me to identify or develop alternative, and hopefully, safer and more effective processes. There were many others who assisted with this book by reviewing, commenting and offering their welcome criticisms and opinions. To all of them, my thanks.

This book is dedicated, with gratitude, to all the bomb technicians who have the cold-blooded courage to take the "long walk".

Donald S. Williams, CPP RSecP MIExpE MIAPS
JUNE 2016

TABLE OF CONTENTS

Chapter 1

Introduction and Explosive Effects

Bombs are literally devastating. Large amounts of energy are released in milliseconds, creating forces capable of destroying buildings, damaging equipment, killing and injuring people. Even the threat of a bomb can cause fear and disruption.

Bombs, also known as "improvised explosive devices" or IEDs, have been used to commit crimes such as murder and insurance fraud, for industrial sabotage, to gain access to valuable assets, to intimidate, and as the primary tool for insurgents, revolutionaries and terrorists.

While distance provides protection from the effects of an explosion, unnecessary evacuations cause disruption, concern and create additional risks. Moving tens, hundreds or even thousands of people in ways they are not used to, forcing them to use evacuation routes and exposing them to external elements when there is no real hazard, is neither safe nor sensible.

Managers can develop and implement bomb safety and security measures to protect people, information and capability while minimising disruption.

How to use this book

The intent of this book is to help managers understand the issues related to bombs, threats, unattended items, post-blast scenarios and related topics. This book is designed to assist managers to 'think' their way through the problems. It will equip them for both the quiet times when there is the opportunity to plan and prepare and also during the incident when an understanding of the key factors will help in making the best decisions under stressful conditions. The manager is provided with the knowledge to draft and implement a Bomb Incident Management plan appropriate to their specific operating environment.

The book is structured so that:
- Chapters 1 and 2 set the scene and provide background knowledge on types of bombs, why they are used and the effects of an explosion as well as describing the various types of bomb incidents a manager may encounter.
- Chapters 3 to 5 address the planning, preparation and response options for the four types of bomb incidents:
  - Bombs of various types,
  - Bomb Threats,
  - Unattended items, and
  - Post-blast.

- Chapters 7 to 10 provides additional information related to bomb incidents:
  - Hazardous Mail,
  - Search management, coordination and planning,
  - Emergency management considerations, and
  - Training, testing and selection of consultants.
- Chapters 11 to 13 provide additional technical and managerial information:
  - Assessing risks relating to bomb incidents,
  - Physical protective security considerations, and
  - The modeling of explosive effects.
- The Annexes provide detail on:
  - The key points from the book,
  - An outline index for a Bomb Incident Management Plan,
  - Real world examples where bomb incident management has been applied,
  - A glossary of terms used throughout the book, and
  - The bibliography and a suggested additional reading list.

Checklists are provided to assist managers develop appropriate plans and procedures prior to and during a bomb incident.

While there is a logical flow throughout the book, each chapter can be read independently and as a result there is some repetition of themes and concepts that relate to more than one type of incident or management concept.

With knowledge of the factors involved, managers can develop procedures and capabilities that match and support their operating environment, protect their organization and minimise disruption and increase the image of a safe and secure workplace. The better prepared the organization, the safer the staff and visitors and the less disruption to operations.

Managers can also use the guidance in this book to assess existing emergency and security plans to determine if they provide an adequate level of protection from bomb incidents.

Technical information relating to explosive effects is included where appropriate but is not essential to developing the skills required to manage bomb incidents.

Bomb Safety and Security: A Managerial Responsibility

Bomb incidents are a management-level responsibility requiring decisions to be made with limited information and within a limited time. Development of an appropriate and applicable Bomb Incident Management plan based on sound security principles enables the implementation of an effective capability to manage the full range of bomb incidents.

The initial consideration may be that it is preferable to evacuate whenever a threat is received or an unattended item is discovered, as it is better to be "safe than sorry". But, organizations require the ability to continue operating until it is determined that a hazard probably exists and that there is a legitimate need to evacuate all or part of the site. It may be argued that evacuating people and closing down the business processes just because a threat has been received or an unattended item has been found is neither safe nor sensible.

Prior consideration of bomb incident management factors and the creation, implementation and exercising of a Bomb Incident Management plan will make it easier to protect from, identify, respond to and recover from the full range of bomb incidents.

Bomb security builds on common security and resilience principles such as access control, defence-in-depth, security awareness training, business continuity, emergency management, and support for victims.

An effective bomb incident management capability will not only protect the organization but also provide confidence to staff, owners and other stakeholders that their safety has been considered and measures are in place to protect them should an incident occur.

Any individual may receive a bomb threat or see an item that raises concern. How an organization **manages** a bomb incident will determine its reputation as a competent, capable body. It may also determine the organization's survival, including its ability to defend itself in a court of law.

The fundamentals for managing bomb incidents are:
- An understanding of bombs, their effects and why they are used.
- An understanding of the different types of bomb incidents.
- The application of basic security practices to prevent bomb incidents, as far as is possible.
- Consideration of the factors related to bomb incidents in relation to the organization.
- Application of the principles for determining if a hazard may exist and for selecting the most appropriate response.
- Drafting, implementing, practicing and on-going reviewing of a Bomb Incident Management plan.
- Integration of Risk Management, Emergency Management, Business Continuity/ Resilience, Human Resources, training and other management disciplines to provide a sound bomb incident management capability.

Most organizations have some form of Emergency Control Organization (ECO) with one person responsible for managing any emergency that may arise. This person may be the Emergency Manager, Incident Controller, Chief Warden, Security Manager or have some other title. In this book the term "Emergency Manager" is used.

Caveat

While every effort has been taken to provide appropriate and applicable guidance in this book, the onus for making the decisions must rest with the managers at the time. The processes developed must reflect the natural, built and operating environment of the site.

Types of Bomb Incidents

There are four types of bomb incidents that an organization may experience. Accurate and consistent terminology is essential to avoid confusion. In this book the following terms are used:

- **Bomb**. An explosive or incendiary device designed to create damage and injury. A bomb can be made from commercial, military or improvised/home-made explosives and components. A bomb can be hand-delivered, vehicle-borne, part of a suicide attack, projected by a weapon, or delivered to the target by a range of other means. The technical term is Improvised Explosive Device (IED) defined by NATO as "A device placed or fabricated in an improvised manner incorporating destructive, lethal, noxious, pyrotechnic or incendiary chemicals and designed to destroy, incapacitate, harass or distract. It may incorporate military stores, but is normally devised from non-military components."[1] The term is extended to include vehicle-borne IED (VBIED) and person-borne IED (PBIED). Most National Bomb Data Centres or equivalent organizations use variations of this definition.

- **Unattended Item**. An item whose presence is not readily explained and which could contain a hazard such as a bomb.

- **Bomb Threat**. A threat that a bomb has or will be used against the organization or person.

- **Post-blast**. The scene after a bomb explodes, often termed a 'bombing'[2], the term post-blast clearly delineates a bomb incident after an explosion.

Related definitions are:

- **Mail bomb**. An explosive or incendiary device sent through the postal or courier systems. As the delivery, identification, assessment and response options differ from bombs placed on the site by the perpetrator, they have different security and response considerations.

- **Hoax**. An item or threat which does not actually represent a hazard but is designed to

---

1   NATO AAP 6-2007 p 130

2   A bombing is "An incident involving the use of one or more improvised explosive devices (IED) which has functioned. Military explosive ordnance which may not be improvised but which has been used in an illegal manner is also included in this definition", Australian Bomb Data Centre definition

create the impression that there is a real bomb on-site[3]. The term 'Hoax' should not be used until after the incident is concluded and the incident report is being written. The use of the term 'bomb hoaxes' suggests that the subject item or call is already defined as a hoax before any evaluation has been conducted. An item can only be designated a hoax after the threat has been assessed or after an item has been inspected.

- **Secondary hazards.** Those materials on-site that are safe until acted upon by an explosion. Managers need to know the type and location of all hazardous material and processes so emergency services can be briefed. Such information may be held in HAZMAT plans. Some secondary hazards, such as high-pressure oxygen or water lines, may not be classified as hazardous material but can still be secondary hazards if they were damaged by an explosion. Note, not all secondary hazards are stationary, some like fuel trailers can be mobile and some may be temporarily on site.

A complete glossary of terms is provided in Annex D.

A Brief History

Possibly the earliest recorded instance of a 'modern' bomb attack was by Felice Orsini in 1858, who used a sophisticated clockwork device, manufactured in England and transported internationally to France, in an attempt to assassinate Emperor Napoleon III. The device failed and Orsini was captured. In 1881 Alexander II of Russia was killed by a suicide bomber. As this was at least the third attempt to kill Alexander II using bombs, it appears the bomber stayed with the bomb to ensure it achieved the aim.

Explosives have been used to commit crimes and promote political causes since at least the mid-1800s. Irish revolutionaries conducted a series of explosive attacks in the 19th century including using a large bomb against the wall of the Clerkwell prison in London on 13 December 1867. This bomb consisted of an estimated 548 pounds of high grade gunpowder and resulted in 12 fatalities and 120 injuries [4].

A significant change in the ability to use explosives for criminal purposes was the development of packaged explosives by Alfred Nobel in 1866 when he used kieselguhr to absorb nitro-glycerin resulting in 'dynamite'. While a boon to the mining and construction industries dynamite also made criminal bombings easier and safer. Soon after the release of dynamite, Irish revolutionaries became known as "Dynamitards" in recognition of their use of the new product [5].

The Anarchists and Nihilists of the late 19th and early 20th century, along with ethno-nationalists used bombs to assassinate senior government members and functionaries, and to create terror in the public. The Anarchists and Nihilists waged a global campaign of

---

3   If an item is believed to be a bomb it is beyond the capability of an organization to determine if it contains real explosives, this is the role of the investigating emergency services who will examine the contents once the item has been dismantled.

4   Campbell 2002

5   ibid

bombings and shootings. The assassination of Archduke Ferdinand by the Serbian "Black Hand" organization, which may have combined elements of ethno-nationalism with Anarchism, precipitated World War One. They also used converted military explosive ordnance by fitting burning fuses to explosive-filled cannon balls, a technique that has been modernized as artillery shells are modified for remote detonation.

'Black Hand' was also a term used by extortionists in America and particularly the crime gangs of Chicago where between 1900 and 1930 when more than 800 bombs were detonated in Chicago, with 157 explosions in one particularly intense 16 month period [6].

Bombings by politically motivated and criminal elements were so common in the late nineteenth century that Isobel Burton, wife of Sir Richard Burton, the British Ambassador to Trieste and famous explorer, observed in 1881 "If the Italians hosted a party the Bulgarians would bomb it. There were bombs in the markets, there were bombs in the streets, there were even bombs in the sausages. It was at times not pleasant."

Vehicle bombs too, have a long history. The previously mentioned Clerkwell prison breakout involved a wagon but the first large vehicle bomb was probably the wagon containing a dynamite charge and approximately 220 kg of iron weights which was detonated on Wall St, New York on 16 September 1920 killing 38 people. The bombing appears to have been an attack on the J.P. Morgan bank and the 'Robber Barons' of the US industrial age.

Bombings continued during the inter-war period to support political and criminal goals. During World War Two, bombs in various forms were used by both sides militarily, and particularly by civilian and paramilitary resistance groups. After WWII, bombs were a main weapon in the post-colonial independence fights around the world. All modern insurgencies and wars of independence have included the use of bombing campaigns as means of both destroying the infrastructure and war-fighting capacity of the regime and to create terror and uncertainty in the minds of the local and international audiences. Algeria, Israel, Afghanistan, Iraq and many South American countries are examples.

Experience and statistical analysis show that bombings are still a preferred weapon of terrorists as well as being a common tool for criminals[7]. For additional reading see the bibliography at the end of the book.

Elements Required for a Bombing

In order to apply bomb security management principles, it is necessary to understand the elements required for a perpetrator to conduct a bombing. These are motive, material, knowledge and opportunity.

---

6   Russo 2001

7   This is supported by information from National Bomb Data Centres (BDC) which collate and publish statistics in classified or unclassified formats on the use of explosive devices.

- **Motive**. The motives for using or threatening to use a bomb may be criminal (murder, extortion, intimidation, vandalism etc.); political (issue motivated/cultural/terrorism etc.); or personal (mental illness, disgruntled employee or client, domestic dispute etc.). As a result, any site may be the target for a perpetrator with a motive against the organization or against someone on the site. A review of possible motives may provide a basis for determining what sort of attack may be expected and why it may be launched. Such an assessment is likely to be generic, but may provide a basis upon which to develop more specific reviews [8].

- **Material**. A bomb requires a main charge of explosive or incendiary material; an initiator to detonate the explosive or ignite the incendiary; a triggering mechanism; and a safety switch [9]. The explosive or incendiary material and components can be commercial, military or homemade. It should be acknowledged that the material to manufacture a bomb exists and is available.

- **Knowledge**. Knowledge to build a bomb may be obtained through formal training in the use of explosives provided to the mining, rural and construction industries; pyrotechnic and special effects companies; and military and law enforcement agencies. In addition, knowledge may be gained through personal research of texts and over the internet. Terrorist organizations provide formal training in manufacture and use of bombs. To successfully place a device on site the perpetrator requires knowledge of the organization, its layout, security and procedures. Little knowledge is required to make a bomb threat.

- **Opportunity**. The perpetrator requires the opportunity to place the bomb on or near the targeted site or individual. This is the element which an organization can control through its security measures. It is possible to deny access to the perpetrator to some areas; limit the ability to bring explosive devices onto the site through the use of detection systems; and to have in place the ability to detect and respond appropriately to bomb incidents.

Advantages and Disadvantages to the Bomber

The use of a bomb provides the perpetrator with certain advantages compared to other forms of physical attacks. It also offers disadvantages which work to the benefit of the targeted organization.

Advantages to the Bomber

Potential advantages to the perpetrator include:

- Large amount of damage created compared to the size of the device.
- Greater level of damage than from an armed assault.

8   See Williams D "Using a study of motives to determine the type of bomb that may be used" October 2003

9   The safety switch allows the bomber to build and move the bomb without it detonating.

- Degree of anonymity as the perpetrator does not have to be on site at the time of the explosion[10].

- If the asset is on or near the boundary then the perpetrator may be able to attack the asset without having to enter the site. Bombs offer the distinct advantage of creating considerable damage from outside a fence. Noting that the greater the distance of the asset from the fence the larger the bomb that has to be manufactured and deployed.

- The effects of a bomb threat may be considerable, including disruption to the target organization's operation and reputation with little expense incurred by the perpetrator, hence the term 'ten pence terrorism'[11].

- Bombings and bomb incidents tend to have high media value and can be used as a means of publicising a cause.

Disadvantages to the Bomber

Disadvantages to the perpetrator include:

- Possibility of the bomb-maker blowing themselves up during construction or transportation of the bomb.

- Possibility that the bomb may not function as intended or at all.

- Without considerable technical knowledge and experience it is difficult to predict the effects of the explosion making it an inexact and variable weapon.

- Risk to the perpetrator of detection when placing the bomb either on or near the site, or into the mail system.

- There is a risk that the device may be detected before it detonates resulting in an evacuation and possibly allowing the responding bomb squad to disarm it.

- There is considerable forensic evidence from the construction of the bomb, which increases the risk of identification and arrest.

Explosives and Explosive Effects

To effectively plan for bomb incidents it is necessary to have an understanding of the effects of an explosion[12]. The visual effects seen at the movies do not reflect normal explosive events as they either exaggerate the effects of an explosion or show victims walking away with no injuries.

Explosions

Bombs primarily use 'chemical explosions' but they are not the only means of generating

---

10   For suicide bombers the considerations of detection on site do not apply to the same degree as the bomber does not need to be as surreptitious or to have an exit plan. Considerations for managing the threat posed by suicide bombers are included in Chapter 3.

11   A term used by UK law enforcement to reflect the (historical) cost of a local call.

12   The information presented here is a simplistic summary of a complex field of gas and hydro-dynamics. Technical information on blast effects is available from engineering, mining and military texts and from professional organizations.

an explosive effect. The basic types of explosions are:

- A **chemical** explosive, in simple terms, is a chemical composition where the molecular bonds can be relatively easily broken causing the material to become a gas within a very short time. A gas occupies a greater volume than the original solid or liquid explosive and hence forces the surrounding air and material away from the centre of the chemical change. What differentiates an explosive is the rate at which the change from solid/liquid to gas occurs. In a fire the solid material is converted to gas at a very slow rate, in a chemical explosion the rate of change is measured in terms of 1,000's of meters per second.

- A **mechanical** explosion is the result of a build-up of heat and pressure inside a vessel. The pressure may eventually reach a point when it overcomes the structural or material resistance of the vessel and an explosion occurs, for example a bursting water boiler.

- A **gas** explosion is where an increase in pressure causes a mechanical rupture or a cloud of combustible gas ignites. Gas explosions are further defined as: Boiling Liquid Expanding Vapour Explosion (BLEVE), Vapour Cloud Explosions (VCE) and a" "Jet" of burning gas. A VCE is where a cloud of gas reaches a critical fuel air balance, referred to as the 'stoichiometric' range, and comes into contact with an ignition source. In a VCE it is possible for the burning cloud front to achieve a 'deflagration to detonation transition' resulting in an expanding high-pressure, high velocity detonation front[13].

- A **nuclear** explosion results from either the fission or fusion of atomic nuclei under extraordinary pressure. Nuclear events are orders of magnitude greater than any conventional explosion.

Explosives

Chemical explosives are roughly grouped into 'High' and 'Low' explosives, delineated by the Velocity of Detonation (VoD) or the speed of the chemical change over a linear distance. Low explosives have a slower VoD and the change from a solid to a gas passes through a combustion stage. Examples of Low explosives include: propellants e.g. gun powders, smokeless powders and black powders used in small arms cartridges; pyrotechnic/firework compositions; and special effects systems as used in the film industry. High explosives generate and sustain a detonation wave where the chemical change occurs without progressing through the burn stage, rather the material is converted into a gas at a rate greater than ~4,500 meters per second. Examples of High explosives are mining explosives, military grade explosives used in bursting artillery shells, explosives used to cut and shape metal, and explosives used to demolish buildings.

---

13   See, for example, CCPS 1994

The effective difference is that Low explosives are suitable for propelling where High explosives are used for shattering and moving greater loads. The shattering effect is known as 'brisance' and is related to the VoD of the explosive.

Explosives can be compared to each other. The standard measure is TNT (trinitrotoluene) which has a VoD of 6,900 meters per second and 1 kg of TNT will produce about 730 liters of gas. TNT is given a sensitivity rating of 1, as a comparison Ammonium Nitrate/ Fuel Oil has a TNT equivalency of ~0.8.

Explosive Effects

Depending on a bomb's location the blast effects may injure and kill people, damage sensitive IT equipment, cause various levels of structural damage, remove sprinkler heads and damage water pipes, disrupt building services including communications and cause structural failure resulting in additional casualties and hindering rescue and emergency responses.

The three primary products of an explosion are blast, fragmentation and heat.

Blast

The gas released from the rapid breakdown of the explosive material requires considerably more volume than the initial explosive occupies, so it moves rapidly away from the centre of the explosion. Initially the gas expands at supersonic speeds but quickly reduces to the speed of sound. The rapid expansion of gas can be thought of as a wall of compressed air, the blast wave, traveling at close to the speed of sound, the front edge of which is the 'shock front'.

Immediately behind the blast wave is a low pressure area from which the compressed air has been drawn and this low pressure area, sometimes incorrectly referred to as a 'vacuum', further damages structures weakened by the blast wave by causing them to stress in the opposite direction.

Blast is applied to an object in two forms, first directly through the sudden increase in pressure referred to as the 'Peak Incident Pressure'. Second if the object withstands the blast, pressure builds up against it until either the object fails or the pressure is reflected; this is referred to as the 'Peak Reflected Pressure'. A limited but viable analogy is the way in which a wall of water builds up against a seawall before being reflected back.

A key factor is the impulse, the time over which the pressure is applied. In relation to High explosives this is tens to hundreds of kilopascals applied over milliseconds resulting in the explosive damage to structures and people[14]. For Low explosives the pressures are lower and slower but can still result in damage and injuries.

14   See Yates 2012

The nature of the blast wave is dependent upon the following variables:

- The type and method of detonation of the explosive. A military grade high explosive has different work properties to an agricultural explosive such as an ammonium nitrate and fuel oil mixture (ANFO). A high-brisant explosive is one in which the maximum pressure is attained so rapidly that the effect is to shatter any material in contact with it and all surrounding material. Low-brisant explosives are of greater benefit where the intent is to 'push' the material after breaking it for example in mining operations.

- The location of the explosion in relation to reflecting surfaces which can increase the effect of the blast wave.

- The proximity of combustible materials and other secondary hazards could augment the effects of the explosion.

- The manner in which the explosive is confined. A greater pressure is required to cause the release from a container therefore the initial blast effect will be higher than for an unconfined explosion.

Figure 1 is a standard pressure/time (pt) curve for an unconfined explosion. The pressure rises from the ambient pressure to its peak within milliseconds as the shock front and blast wave pass over the observer. After the blast wave passes, the low-pressure phase then occurs. The area under the pressure curve represents the impulse when the pressure is applied against any resistant structure for a period of time.

The time/pressure curve shown in Figure 1 is for an idealised hemispherical blast wave. For blasts surrounded by buildings, the graph will be radically different, with repeated waves of high and low pressure appearing as the shock wave and pressures reflect off surfaces.

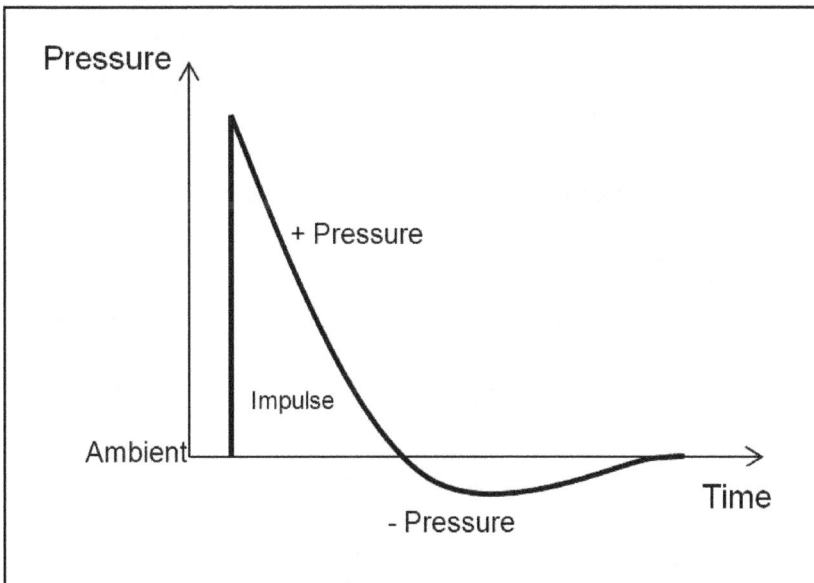

Figure 1 – Idealised Time / Pressure curve

The blast wave dissipates quickly based on a cube-root[15] scale therefore the greater the distance that can be provided between the location of the bomb and the target the greater the protection from bomb effects. Distance is of significant benefit when developing bomb incident preventative and response measures.

Fragmentation

The casing of the bomb and anything nearby will be shattered and turned into high velocity projected fragments. In some cases the bomb is designed to enhance the fragmentation damage through the addition of nails etc. The lethality of the fragments depends upon their mass and velocity.

Fragments travel away from the initiating site of the explosion in an approximation of a straight line before being brought to ground by gravity. Therefore, the likelihood of being struck by a fragment is related to distance. Protective structures will also reduce the likelihood of damage from fragmentation strike.

A major consideration in relation to blast effects is the projection of glass fragments from external glazing and internal glass partitions. Glass fragments can be responsible for a large number of lacerations, penetration injuries and particularly eye injuries[16]. Also to be considered are nearby fittings, street furniture, and other objects which may become 'secondary' projectiles with a separate set of assumptions and ballistic properties.

Fragments can be expected to travel further than the blast effects and subject people and structures, to impact damage.

Heat

The detonation process is exothermic, generating considerable heat. 1 kg of TNT will generate a temperature of approximately 1000 K cal/kg (730°C). For many explosives this heat does not last long and it is unlikely a fire will start unless there are combustibles in the immediate area or the bomb has been constructed with accelerants to enhance the incendiary effect. Some bombs are specifically designed as incendiaries[17].

Summary

The use of bombs for criminal and political purposes is not a new phenomenon.

Bomb incidents offer considerable benefits but also significant disadvantages to the perpetrator.

The effects of an explosion are achieved in milliseconds when the sudden conversion

---

15   For a hemispherical blast wave i.e. one detonated at or near ground level.

16   See Chapter 12, Physical Protection, for observations on glazing protection

17   For additional information on blast effects refer to Yates 2012.

of the explosive material to a gas creates an expanding blast wave, intense heat and fragmentation. The impact on people, structures and infrastructure can be severe.

To undertake a bombing the perpetrator requires the four elements of: motive, material, knowledge and opportunity. Understanding this assists the manager in planning how to limit the opportunity for a perpetrator to successfully attack the site.

Bomb incidents are a management-level responsibility requiring decisions be made with limited information and within a limited time. Development of an appropriate and applicable Bomb Incident Management plan, based on sound security principles provides an effective capability to manage the full range of bomb incidents.

The fundamental principle of bomb security is the ability to keep the bomb some distance from the asset and the ability to move the assets, particularly people, away from the bomb.

With prior planning and preparation an organization can manage bomb incidents thereby providing a safer and more effective workplace.

There are four types of bomb incidents for which managers require plans:
- bombs of various types,
- threats,
- unattended items, and
- post-blast.

Chapter 2

Bomb Security Planning

Bomb Incident Management Plan

A bomb incident occurs quickly and requires a managerial response based on limited information.

Each organization should create a Bomb Incident Management Plan. The aims of the plan should be to:

- provide the information, policies, procedures and training to limit the ability of a perpetrator to cause a bomb incident on the site, and
- enable the organization to respond in the safest, most effective and efficient manner if an incident should occur.

A Bomb Incident Management Plan enables an organization to prepare the procedures, training, people, equipment and communication lines to reduce the likelihood of a bomb incident and to minimise losses should an incident occur. The plan should provide the information in a clearly defined and documented manner.

The format of the plan should provide specific guidance, policies and procedures for fire wardens, security supervisors and relevant managers. The plan should address how staff, on-site contractors, tenants and visitors will be informed of the correct response. The plan will need to be site-specific, as each location will have a particular operating environment, assets, resources and constraints.

For most organizations the likelihood of bomb incidents are as per Figure 2.

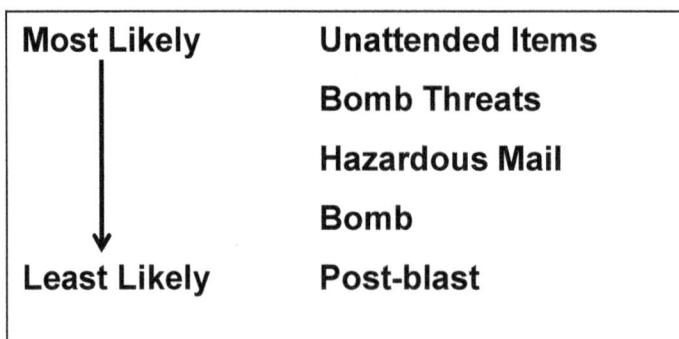

Most Likely	Unattended Items
↓	Bomb Threats
	Hazardous Mail
	Bomb
Least Likely	Post-blast

Figure 2 –Bomb incidents - likelihood rating

An organization's Bomb Incident Management plan should address all of the above. An example index for a Bomb Incident Management plan is provided in Annex B.

Bomb Security and Safety Principles

For a Bomb Incident Management Plan to be effective it should be based on:

- An understanding of the elements of a bombing: motive, material, knowledge and opportunity.
- The exposure of the organization to bomb incidents due to its business environment, location, relationships and staff.
- The effectiveness of related emergency, health and safety, HR and other plans.
- Current protective security measures.
- Commitment by senior management to recognise that the organization may be subjected to bomb incidents and a specific Bomb Incident Management plan is required.

The principles for bomb security are:

- **Preventing a bomb from entering the site.** This may be achieved through access controls and screening of visitors and goods, including mail. Where strict access control is not possible access to critical areas on the site should be controlled [18].
- **Early detection of bomb incidents.** This will rely on trained and aware staff with clear procedures for reporting of items and threats.
- **Appropriate response measures.** This will be based on clear and practiced procedures for evaluating the potential impact of bomb incidents and applying the relevant responses to protect people and the organization's assets and reputation.
- **Design of facilities.** To reduce the likelihood of a bomb being brought onto the site to increase the likelihood of detecting a bomb to mitigate the effects of an explosion.

Factors that will influence the ability of the organization to protect itself effectively and appropriately from a bomb incident include:

- The existing social, political, workplace and related activities, which might make the organization a target for an act of violence. The likelihood of a bomb incident will alter over time.
- Basic access control measures including adequate boundary protection, personal identification and verification systems, electronic access control and recording systems, key control processes, restricting the public-to-public areas etc.
- The security and control measures put in place by neighbouring businesses, particularly in multi-tenanted sites.
- "Defence in Depth", providing additional levels of control and restricted access to higher value assets.
- Good workplace practices such as keeping areas tidy and clean so that any items

18 See Chapter 12, Physical Protection

introduced to the work area will be quickly identified.

- Staff awareness and an ability to identify items which are out of place and to report them to their supervisors/managers.
- Supervisors/managers having the training and knowledge to respond appropriately to a report of a bomb incident from staff or the public.
- The proximity of assets to the boundary e.g. where the building housing critical assets or functions forms part of the boundary and the bomb can be placed external to the site. In such cases the likelihood of detecting such an item and the ability to respond to it should be assessed. CCTV and mobile guards may be factors in this assessment.
- If bomb detection equipment has been deployed, it is important to determine if it is appropriate and capable of the task for which it is employed; that the equipment is part of a cohesive security plan; that it is maintained; that staff are trained; and that bomb incident management processes are in place. A key question is "What will we do when we find what we are looking for?"

The above factors are those required to maintain a normal safe and secure work area. Managing bomb incidents should build on existing security and safety measures.

The Bomb Incident Management plan should describe how bomb security training will be developed and delivered, in particular, training on the evaluation of bomb threats and unattended items and in search techniques.

Of course, managers always have the option of evacuating and closing down the organization's operations if they believe this is the safest and most appropriate option.

Real-world examples of how bomb incidents have been managed are provided in Annex C.

Related Plans

The existing security measures at the site all influence bomb security planning. These include the ability to control access to all or some of the areas on the site; the inspection of people and vehicles entering the site; the inspection of mail; the amount of CCTV coverage and the ability to quickly access the images; and related security measures. Bomb security, similar to most aspects of security, cannot happen in isolation. It requires the support and interaction of a range of business units. Bomb security planning will require input from: Security, Emergency Management, Facility Management, Human Resources, Workplace Health and Safety, Environmental Management, Business Operations, Media Management/PR, Legal and Finance.

The Bomb Incident Management Plan should ensure that the related plans are consistent in their requirements. The other plans may need to be modified as bomb-specific considerations are identified, such as evacuation distances and business disruption due to crime scene considerations.

Information Gathering

For effective bomb security analysis, information is required on:

- The level and type of 'normal' bombing activity within the area of interest.
- Detailed knowledge of the site and activity under review.
- Trends and changes in bomb incidents.
- Political and issue motivated groups and their motives and methods of operations.
- Specific people or sites which are or may be considered targets.
- The availability of explosive materials or their constituents.
- Planned or possible changes to the working and political environment.

Bomb incident information may be gained from:

- Open source media.
- Historical records.
- Organizational data.
- Professional organizations.
- Intelligence agencies.
- National Bomb Data Centres (or equivalents).
- Computer modeling.
- Corporate, national and international liaison.

Assessing the Risks

Appropriate measures to provide security from, and safe responses to, bomb incidents may be developed using security risk management principles and techniques to identify the likelihood of such incidents, and the potential consequences should the incident occur. See Chapter 11, Risk Assessment for details on assessment and mitigation measures for bomb incidents.

Summary

Every organization should have a Bomb Incident Management Plan that addresses how it will deter, recognise and respond to: bombs of various types, threats, unattended items and post-blast incidents.

The plans should be based on a sound basis of bomb-security measures, local knowledge and available intelligence as to the threats.

The Bomb Incident Management Plan should be integrated with related security, safety, emergency and other corporate plans and should augment rather than conflict with the existing plans.

Managers always have the option of evacuating if they believe this is the safest and most appropriate option.

Chapter 3

Preparing for and Responding to Bombs

Even though the likelihood of being bombed may be assessed as an unlikely security incident, the consequences of an explosion can be catastrophic and appropriate preventative and mitigation measures should be planned and implemented.

The principles outlined in this Chapter apply to all types of bombs: hand-delivered, vehicle borne, mail bombs[19], suicide bombs, etc. The detailed application of the principles will vary depending on the threat analysis, exposure and vulnerabilities of the organization.

Consequences should a bomb enter the site include:

- An explosion could occur, in which case the specific consequences will vary depending on the construction, type and location of the bomb, particularly its proximity to various assets. Response considerations for an explosion are provided in Chapter 6 "Post-blast".

- There will be disruption to operations, should the bomb be identified, while an evacuation is initiated and the incident is managed by the site's Emergency Control Organization[20]. The site may be unusable for hours, or after an explosion possibly for weeks or permanently.

- Staff, clients and public may be concerned over the manner in which the incident was managed. Documented and rehearsed procedures will assist with this issue.

- Hazards will be faced by people during an evacuation such as road crossings, handicapped egress etc. these should be addressed in the site's emergency procedures.

The primary treatments for responding to a bomb include the ability to detect the item and to respond appropriately and safely. The outline process for assessing and responding to a bomb on site is shown in Figure 3.

Identifying a Bomb

Staff, including security staff, are the most likely to identify a bomb, as they are the ones familiar with the work environment. In some cases members of the public may report items they believe are of concern[21].

Identification of a bomb is based on noticing that which is out of place and does not fit its environment. In some cases the item will be immediately identified as a bomb due to its construction, location or because it is related to a threat, forced entry, or similar indicator.

---

19   Additional considerations for hazardous mail items are provided in Chapter 7.

20   See Chapter 9, Emergency Management

21   An item believed to be a bomb can only be classified as a "Hoax' (i.e. not containing real explosive components) by the responding emergency services personnel or through subsequent forensic examination. For assessment of Unidentified Items see Chapter 5.

Because bombs are 'improvised' it is not possible to predict what they will look like or where they will be placed. Security measures may reduce the likelihood of a perpetrator trying to get a bomb onto the site and increase the chances of finding a bomb before it explodes. These measures can include having access control measures to limit where the public can go; mail screening; and aware and trained staff. For physical security measures to protect against and prevent bombs from entering the site see Chapter 12, Physical Security.

Types of Bombs

There are many ways in which bombs can be delivered to the site and concealed both in transit and once delivered. Understanding the different methods by which a bomb can be delivered and the use of consistent terminology will assist in both planning for and responding to the various types of bombs.

Person Borne or Placed Bombs

Person borne IEDs (PBIED) or placed bombs are the most common form of bomb, where the perpetrator carries the bomb onto the site places it and departs. The size of the bomb can be determined by considering that a 5 kg (11 lb) weight can be held with an outstretched arm, 10 kg (22 lb) can be carried by the side of the body, 20 kg (44 lb) is a heavy two-arm carry and anything above that will be transported on wheels. The maximum weight for most airline hold luggage is 23 kg (50 lb) and this is a heavy bag on wheels. If the site is one where large luggage items may be part of the normal operating environment then planning for a bomb of 23 kg may be required. Otherwise, most placed bombs are likely to be in the 5 to 10 kg (11 to 22 lb) range.

Ambush Bombs, Secondary and Multiple Devices

It is possible that more than one bomb may have been placed on or near the site. This is another case where terminology is important if confusion is to be avoided. The following definitions relate to the motive of the perpetrator and are used in this book:

- Ambush Bombs. Devices placed along exit routes, assembly or other areas with the intent of killing those leaving the site or passing the bomb.

- Secondary Devices. Placed to attack the responding emergency services. Usually deployed by groups that have decided that the emergency services, including bomb squads, are defeating their efforts and should be targeted. Placement of secondary devices usually requires the perpetrators to predict how the responding emergency services will deploy and where.

- Multiple Devices. Where the perpetrator places more than one bomb at the site. The intent being to cause maximum damage through multiple simultaneous, or near simultaneous explosions.

Body Bombs

A 'body bomb' is a sub-set of a PBIED, which is worn by the perpetrator. It may be a suicide bomb or it may be fitted to a victim particularly as a negotiating tool for a robbery, extortion or kidnapping. Body bombs can be used by those that are mentally unstable to attract attention. Body bombs or the threat of body bombs have been used as a weapon during theft, particularly from banks. Suicide bombs have been recorded since the late 1800's but the resurgence in the late 20th and early 21st Centuries has raised the awareness of this type of attack[22].

Organizations should conduct a formal threat assessment to determine if they are likely to be the target of a suicide bomber, who represents a very small sub-set of bombing perpetrators. Considerable research has been conducted into the motives and attack methods of suicide bombers, which may inform security managers concerned about this type of attack.

The management of suicide bombers requires similar considerations to other types of bombs: identifying the hazard and having responses in place. The identification and response to suicide bombers may be more difficult as there may be less time to respond and the suicide bomber is the "ultimate guided weapon", capable of changing its target and timing[23].

If a suspected body bomb is identified, usually the best way to minimise the consequences is to immediately begin to move people away and to limit the bomber's ability to get closer to the organization's assets. History shows that once a suicide bomber believes they have been detected or is to be thwarted in their attack they will try to detonate the bomb to cause as much effect as possible. If a suicide bomber does detonate the bomb a post-blast plan will be required[24].

Vehicle Bombs

Vehicles provide the perpetrator with the advantages of being able to enter the targeted site with the bomb already prepared thereby reducing the time the perpetrator has to spend on site. They can be used to break through barriers, particularly if driven by a suicide bomber; they tend to blend into the environment making them difficult to identify as unusual or out of place; many sites have public parking in their sites; and vehicles can carry large quantities of explosives. It is stressed that not all vehicle-borne IED (VBIED) are suicide attacks. The 1993 bombing of the New York World Trade Centre, the bombing of the Edward P Murrah Building in Oklahoma City and the attack against the USAF Kohbar Tower in Saudi Arabia are examples where the drivers left the vehicle.

---

22   See Williams C. 2004, Bloom M. 2005, Pape R. 2005

23   Williams C., 2005 "Suicide Bombers" presentation to Australian Homeland Security Research Centre Conference.

24   See Chapter 6, Post Blast

The detection of a VBIED is difficult, particularly where public parking is provided as part of the organization's operating environment. Detection may depend on identifying unusual behaviour of the driver and passengers. Once detected the vehicle should be investigated to determine if it or the related activity was suspicious, the hazard posed, and the safest and most effective response.

Related to VBIED are bombs placed in or on vehicles to kill the occupant, these are usually smaller devices positioned under the driver's or passenger's seat depending on where the target is expected to be located. Protective considerations against this type of attack include:

- Secure garaging of the vehicle.
- Securing the vehicle at times when it is in a public area.
- Training staff to look for and report signs of disturbance or tampering.
- Searching the vehicle prior to use and after it has been unattended.

Off-route bombs

Bombs can be used to attack vehicles as they travel. Bombs have been buried under bridges[25], placed on the side of the road or fired from a distance using military or improvised rockets. Specially designed shaped charges and explosively formed projectiles, sometimes referred to as 'off-route' mines have also been used to attack vehicles.

The primary protection from off-route is through varying routes, keeping information about travel plans secure, training drivers and escorts to be aware of the hazards, and identifying anything that is out of the ordinary.

Projected bombs

Projected bombs are those that are fired into the target site. They can be improvised mortars, improvised shoulder fired grenades or rockets, or projected using a slingshot or similar tool. While not having the range or accuracy of professionally designed and manufactured military items they do allow the perpetrator to be some distance from the target. Projected bombs tend to explode on impact and (unless they fail to function) offer little opportunity to evacuate or render them safe. The IRA used remote and timed initiators so that the perpetrator did not have to be near the launch point at the time of firing. Having a bomb at the launch site also helped to disrupt the forensic evidence[26].

Dirty Bombs

Bombs may be augmented or contaminated with other hazards such as poisons, biological or radiological material. Biological material is less likely as the intense heat of an explosion

---

25   Such as the assassination of Judge Falcone in Sicily in 1992.

26   Advances in explosive forensic science have meant that useful evidence can be obtained from most explosive crime scenes.

may destroy the material. Poisons and other chemicals have been used to increase the lethality of fragmentation. The possibility of radiological material being added to an explosive device to contaminate an area has been a concern for many years.

It is unlikely that the manager will know that the bomb has been augmented or with what material. It is probable that contamination by chemical, biological or radiological material will only be detected during the post incident investigation.

This is one reason why all those exposed to the item or explosion must have their details recorded; see Chapter 9, Emergency Management.

Mail and Courier-delivered bombs

The mail and courier services have been used to deliver bombs to specific addresses. Specific considerations for identifying and responding to mail and courier-delivered bombs and other hazards are provided in Chapter 7, Hazardous Mail.

Responding to a Bomb

The organization's emphasis should be on protecting people and other assets until the emergency services arrive, or in the worst case, the bomb explodes[27].

Protection from a bomb is achieved through distance and cover. The bomb should not be moved and no attempt should be made to 'disarm' it or make it safe. Only qualified bomb technicians have the training, experience and equipment to carry out 'render safe' procedures.

The planned response should include:

- Reporting the incident to the Emergency Manager or other nominated position.
- Confirming whether the incident is a bomb or an unattended item. If the item is an unattended item then an assessment should be conducted to determine if it may pose a hazard, see Chapter 5, Unattended Items.
- Initiating an immediate evacuation, beginning with those closest to the item. Depending on the size and location of the bomb a decision on whether a full or partial evacuation of the site is required, see Chapter 9, Emergency Management.
- Managing other aspects of the close down and evacuation procedures in accordance with the organization's emergency procedures.

If the bomb is in another part of the site, consideration may be given to keeping people inside, based on 'Shelter-in-Place' procedures. This will depend on the relationship between the buildings, the size and location of the bomb, the structural strength of the building, and similar factors. The decision will be a balance between the exposure

---

27   See Chapter 6, Post-blast

presented by evacuating people and the risks of leaving them in place. If people are kept in buildings they must be moved away from any windows or glass facing the direction of the bomb and should use intervening walls as added protection.

Depending on the motive and skill of the perpetrator there is a possibility of other bombs being on or near the site. The emergency procedures should include the requirement for the egress routes and assembly areas to be inspected for unattended items prior to or during the evacuation. Such an inspection is good practice, as it will also detect blocked exit routes and other hazards.

The organization's evacuation plans should be reviewed to ensure they provide adequate distances and alternative evacuation routes and assembly areas suitable for explosive hazards. The specific response plans will vary by site, organizational function, and assets based on the security assessment.

Where possible the evacuation assembly areas should be:
- At least 100 meters from the building.
- Out of direct line of site of the bomb.
- Not facing, or under, windows.
- Behind solid cover such as another building.
- Not exposed to other hazards.

An emergency rendezvous point (ERP) should be established to meet and brief the responding Emergency Services. The witnesses who have seen the bomb should be available to brief the responding emergency services. The sooner the emergency services can gather the required information, the quicker their response will be and the greater the chance they will be able to disarm the bomb before it explodes and damages the organization's business.

In addition, a manager should be at the ERP who knows the facility and plans of the site.

See Chapter 9 Emergency Management for additional considerations.

After the incident is complete there will be a need for reports of what happened. The reports will be required by the organization and possibly by external agencies investigating the incident. The reports will also provide an ability to review the response actions and to improve policies, procedures and training.

Figure 3 outlines the considerations for responding to a bomb on site.

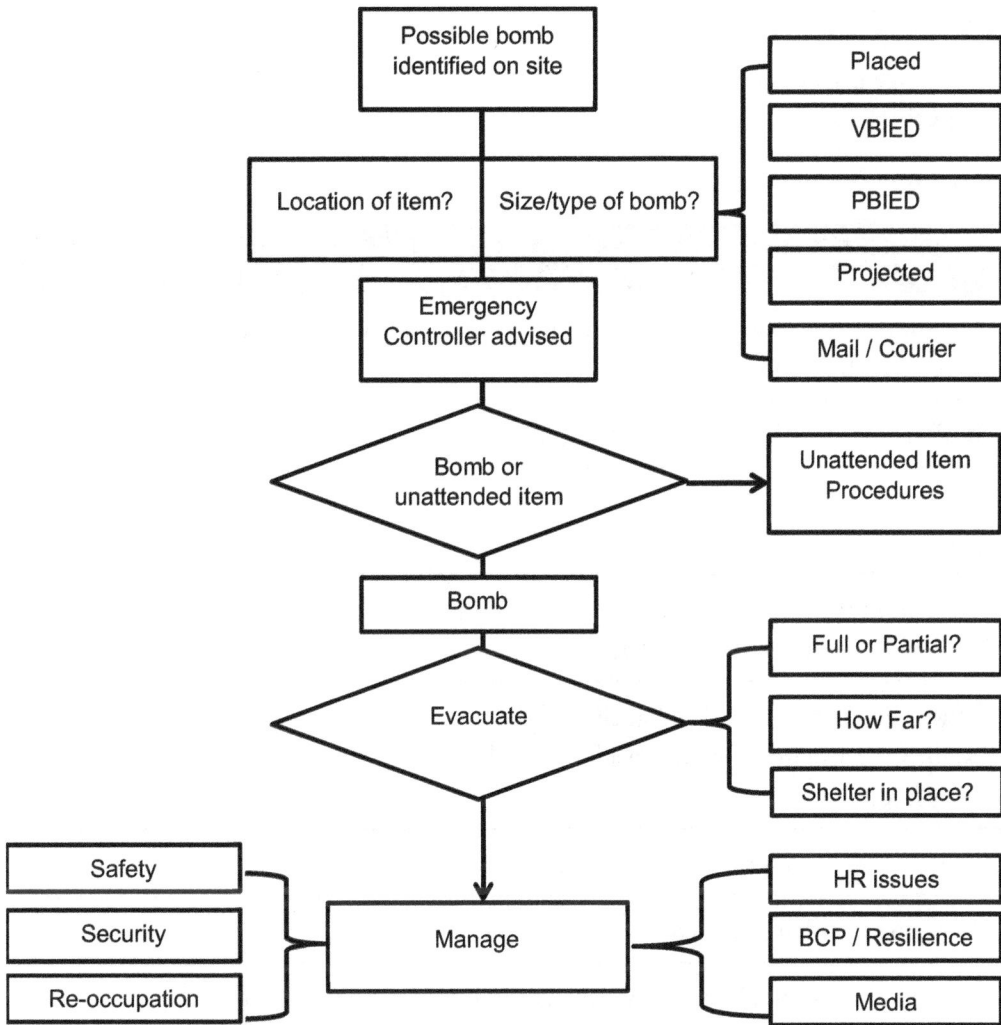

Figure 3 – Suspected bomb on site flow chart

Summary

Bombs can be delivered in a wide range of forms, which will, to some degree be restricted by the security measures of the site.

If an item believed to be a bomb is identified on site managers need a prepared, effective and efficient plan that will protect people and functions while working in alignment with other organizational plans such as the emergency, safety, security, business continuity and media management plans.

The primary response to a bomb is evacuation and placing adequate distance and shelter between the people and the bomb.

Real world examples of bomb incident management are provided in Annex C.

Checklists

The following generic checklists are provided to help organizations develop their own site-specific procedures and immediate response guides.

Planning Considerations

Bomb On Site - Planning	Response
A manager, and deputy, has documented responsibility for responding to bomb incidents.	
Staff know how to identify something that may be a bomb.	
Staff know how to report what they think is a bomb and to whom.	
Evacuation routes and assembly areas – suitable for a bomb incident – have been identified and documented.	
Other Emergency, HR, Media, BCP/Resilience, Security, Facility and Environmental considerations have been addressed.	
Contact lists are appropriate and maintained.	
Reoccupation plan prepared.	
Method of advising all on site of evacuation including direction, assembly area, etc. exists.	

Incident Checklist

Bomb On Site - Action	Response
Item reported to responsible manager when?	
Item confirmed as probable bomb rather than unattended item.	
Identify type of bomb and possible size.	
Emergency Services advised?	
Full or partial evacuation?	

Bomb On Site - Action	Response
If bomb is in another location is shelter in place the safest option?	
Secondary hazards in area?	
Safest evacuation route?	
Safest assembly area?	
Emergency rendezvous point established and Emergency Services advised?	
Staff and visitors evacuated, site checked?	
Hand back from Emergency Services after incident finished?	
Safe reoccupation process?	
Internal and external reports completed and submitted.	

Chapter 4

Preparing for and Responding to Bomb Threats

The threat of violence against an organization can be terrifying; a simple phone call can cause massive disruption and fear.

A 'threat' can be considered to be any claim that an act of violence will be perpetrated against the organization or an individual. This Chapter relates to bomb threats but the threat assessment and response principles can be used to assess any threat received by an organization. The threat evaluation process adopted by an organization should be capable of dealing with the full range of threats.

To label a threat a 'hoax call' or 'hoax threat' before it is evaluated negates the entire process.

Staff, contractors and possibly even visitors, should know that threats are treated seriously by management and that there is a process in place to evaluate them.

Why Threaten?

A key consideration is: why would the perpetrator provide a warning that there is, allegedly, a bomb on the site? A perpetrator wanting to kill is unlikely to provide a warning unless it is to channel people into a killing zone for an ambush device[28]. If they are providing a warning it may be that they wish to damage property but not to kill, or that there has been a change of heart since the bomb was placed. Or, the motive may be to cause disruption and fear in the organization and there is no bomb.

For some organizations the number of bomb threats peak at certain times such as in educational institutions where they tend to be more common at exam time than during holidays. Some industry sectors such as the aviation industry receive many threats in a year and deal with them on an almost routine basis. Other organizations are rarely if ever threatened, it is these that are at greater risk of failing to evaluate and respond in the safest and most appropriate manner.

A structured, documented and practiced threat evaluation capability will reduce the likelihood of failing to respond appropriately, as an improbable threat will not create the required effect, whereas a threat that is plausible will be dealt with in a manner that will reassure staff, visitors, clients and owners that appropriate measures were taken.

Bomb Threat Management Principles

The initial, instinctive reaction whenever a threat is received may be to evacuate although

---

28   See Chapter 3, Bombs for a description of Ambush Devices. See Chapter 9, Emergency Management for considerations relating to evacuations in response to a bomb incident.

this may not be the safest or most sensible action. It also does not reflect security management principles. Constant evacuation may undermine employees', clients' and owners' confidence in management's ability to provide a safe, secure and productive work environment. Constant evacuation will also lead to 'copy-cat' incidents as staff may seek time off from work or outsiders enjoy the prospect of disrupting activities. The perpetrator may learn how the organization reacts and capitalise on routine behaviour. At the other extreme, some organizations have an (unofficial) policy of not responding to threats, as "they are always hoaxes"; this approach may be difficult to justify.

The effectiveness of the response to a threat will depend on the organization's ability to:

- Recognise they have been threatened.
- Capture and report the information about the threat to the relevant authority within the organization.
- Evaluate the threat.
- Respond appropriately.

Threat Evaluation

Threat evaluation is a management decision-making process and is not a simple matter. A process is required that permits the threat to be assessed by those with appropriate training and skills. An effective, efficient and rehearsed threat evaluation capability can be developed. Threat evaluation, including exercising of the response options available, should be practiced on a regular basis.

Threat assessment is often presented in terms of determining if the threat is 'specific' or 'non-specific'[29] which is a method of indicating the amount of knowledge (and commitment) the perpetrator has when making the threat. If someone with knowledge of the site makes a detailed threat then it will be given more credibility as it represents knowledge that may have been gained as the bomb was placed. If the perpetrator has gone to the effort and personal risk of making and deploying a bomb and then wants to warn the organization, then it is probable that the perpetrator will provide information that will assist the organization to evaluate and respond to the threat. Someone with little commitment or knowledge of the site can not be expected to provide specific information about the location, motive or design of the bomb.

An effective threat evaluation process will demonstrate to staff, clients, owners, co-tenants, emergency services and other stakeholders that there is a considered, practised and efficient process in place to protect the organization's assets either by evacuating when necessary or by minimising disruption when the threat is not plausible.

---

29   See Newman 2005

Threat Evaluation Team (TET) and the Threat Coordinator

The question is not "Is the threat real?" as we know that the threat exists; the question to be answered is: **"Can the perpetrator have done that which they claim?"** A coordinated, trained and practiced Threat Evaluation Team (TET) with a working knowledge of threat evaluation principles will provide an effective capability to make the best decision on the information available at the time.

The threat should be evaluated by a team, thereby enabling different areas of knowledge to be brought to the evaluation. The purpose of the TET is to recommend to the team leader the appropriate response options. The final decision should rest with a single person, the team leader. Selection of the team and the threat coordinator (the team leader) is important. The team members should be those with a sound knowledge of the site, its operations and its security. The TET should be limited in size to assist in quick assessment and decision-making. A record keeper or secretary is an important part of the TET.

Each of the company's sites should have its own TET. It is very difficult for someone off-site to have the required local and immediate knowledge of the site's operation and security at the time the threat is being evaluated. The constitution of the TET may vary between the organization's sites depending on the operations of the site and the management structure and responsibilities. Possible members of the TET include the managers responsible for: security, safety, facilities, business continuity/resilience and operations.

The TET should be led by a Threat Coordinator. Selection considerations for the position of Threat Coordinator include:

- Be at a suitable managerial level to make decisions that can be acted upon.
- Be authorised to carry out the threat coordination functions.
- Have the ability to make rational decisions under pressure.
- Have an understanding of the organization's emergency response procedures.
- Be a position known and contactable by all staff either directly or through the security or emergency managers.
- Be contactable at all hours, either through one person being on call or through the position being one that is on-site at all times.
- Be one where the person filling the position is usually on site, not a frequent traveller.
- Have identified, trained and documented deputies that may fulfil the functions.
- Be appropriate to the site, it may be that different positions are nominated at different sites.

The responsibilities of the Threat Coordinator are to call together the TET and to coordinate the threat evaluation. Therefore the Threat Coordinator should have the documented authority to:

- Conduct the initial review of the threat and decide if an evacuation should be initiated immediately.
- If an immediate evacuation is not required, call together the TET.
- Gather together the witnesses, the record of the threat and all other relevant information.

Consideration should also be given to how threats received after hours will be evaluated. This may depend upon the level of activity occurring at the time, which managers with appropriate decision making powers and threat evaluation training are on site or contactable and the ability to access the CCTV, access control and other sources of information.

The Five Rs

There are five phases to threat evaluation, known as the 5 Rs:
- Receipt
- Record
- Report
- Review
- Respond

Receipt

Threats may be received by phone, fax, mail, e-mail, social media, SMS, or even delivered by person. An important element in threat assessment is knowing that the organization has been threatened. There should be a system in place to ensure that all threats are captured. All staff should be educated in what to do so that the person who receives the threat does not decide to ignore it.

Record

The evaluation will rely heavily on the record of the threat made by the person who received it. The TET will be relying on what the person who took the call thought the person who made the call said. National Bomb Data Centres and similar agencies publish Bomb Threat checklists and many of the commercial checklists are copies of these.

The checklist should be widely distributed and easily accessed, having the checklist on the corporate intranet may ensure wide distribution but it may not be easy to find or use when needed in a hurry. It should be able to be scanned or faxed and therefore the use of red and other dark backgrounds should be avoided. It should be simple to use with adequate space for entering information specifically the exact wording of the threat. Some bomb threat checklists are so full of prompts there is little room for the actual threat. Staff can be trained to use another piece of paper, or a keyboard, to write down exactly what is said

Accuracy is of critical importance when recording threat information. Most checklists provide room for recording information about background noises, the voice of the caller etc. The information should be recorded along with any other impressions as soon as the call is completed.

In the case of SMS, e-mail, fax or mail threats, there is a more accurate record to work with but the threat must still be captured and recorded including when and how it was received.

The person who received the threat should not talk about it to others; gossip and rumour about a threat will create unnecessary concern and may lead to a distortion of what they remember.

The person who took the call must remain available to talk to the TET. They should be given a quiet place to sit while they write out in greater detail what they remember about the call.

Advice is often given to the receiver of the call not to hang up, as it may be possible to trace the call. The effectiveness of this depends upon whether the call was from a landline and the switchboard technology and its capabilities; each site should determine if tracing calls is possible once the caller has hung up. If the phone has caller ID then the number from which the threat is made should be recorded. If the phone differentiates between internal and external calls, the origin of the call should be noted. If the call is received on a cell phone, the call log should be checked to see if the number of the caller was recorded.

Report

Management need to nominate the position to which all threats are referred, possibly the Emergency Manager. A position should be selected rather than a person so that deputies and temporary appointments to the position may be trained in the responsibilities.

The threat should be passed to the nominated manager quickly and efficiently. As the threat could be received by anyone, a simple and easily recognisable passage of information needs to be defined. An effective method is to ask staff to inform their local fire warden. Wardens should be advised how to pass on the information to the nominated manager and to begin preparing themselves for an evacuation should one be called.

The Threat Coordinator will be responsible for calling together the TET. The Threat Coordinator will also conduct an initial evaluation of the threat and initiate an evacuation if they believe the threat does not permit time for a detailed evaluation by the TET. Therefore the Threat Coordinator should be of an appropriate managerial level with the documented authority to call other managers together at short notice and to evacuate the site.

Review

The review phase is the most important and difficult aspect of the process. It is stressed that threat evaluation is a managerial decision-making process. Unlike a fire it is not a simple identification of a visible threat but a review of facts to determine if the basis of the threat is reasonable.

The primary evidence to be assessed is the actual threat. The assessment of the threat is largely based on what the person who received the call thinks the caller said; the TET is already working on second hand information.

As well as reviewing the wording of the threat there are a number of other sources of information that will help the evaluators assess the feasibility that the perpetrator has done that which they claim. These may include:

- Review of CCTV records.
- Review of access control records.
- Interviews with staff and others.
- Having IT investigate the source of threats received via e-mail or similar.
- Searches of areas, ranging from employee scans of their work areas to formal searches of nominated areas[30].
- Knowledge of current events that may alter the threat profile of the organization because of domestic, international, industrial, political or social changes.

If the threat is to a public area, then basic security and housekeeping measures help; keep the site tidy, confine the public to the public areas, and don't allow boxes and deliveries to accumulate.

Threat Evaluation Time Calculation

The time available to evaluate the threat may be calculated by subtracting the known time to evacuate the site plus a safety margin from the stated deadline, if provided. To work out when a decision has to be made to evacuate, the following equation may be used:

$TD = Td - (Te + Ts)$ where:

TD is the latest time a decision must be made

Td is the time of stated deadline

Te is the known time to evacuate

Ts is the safety margin

30   See Chapter 8, Search

Example:

- time now is 10.30 am
- a perpetrator calls with a bomb threat and gives a deadline (Td) of 12 noon,
- the time taken to evacuate (Te) site = 20 minutes
- the organization applies a safety margin (Ts) of 15 minutes
- then the time of decision (TD) = 12 pm – (20 min + 15 min = 35 min) = 11.25 am.
- therefore a decision must be made no later than 11.25 am.
- this means that the time between time now and 11:25 is the time available to assess the threat and make the decision: 10.30 to 11.25 = 55 minutes to assess, search and consider before having to make a decision.

$$11.25 \ (TD) = 12.00 \ (Td) - (20 \ (Te) + 15 \ (Ts))$$
12.00 – 35 minutes = 11.25 = TD = latest time a decision can be made
10.30 (time now) to 11.55 = have up to 55 minutes to assess the threat

A decision may be made in less time than that available if enough information is received or found to support the choice to continue working or to evacuate.

If the time available calculation indicates that there is not enough time to safely evacuate the site and conduct a threat evaluation then an immediate decision on whether to evacuate should be made by the Threat Coordinator.

If the perpetrator does not provide a deadline then the Threat Coordinator should evaluate the threat as quickly and efficiently as possible.

In the time available, a large amount of information may be gained and evaluated. As an example:

- Is it feasible, given the existing level of security, access control and activity in the threatened site, for someone to have introduced a bomb into the area?
- Can video surveillance and access control records be quickly accessed to determine if there is any indication of unauthorised access?
- Can searches be conducted to identify any unusual or out-of-place items?
- Can staff in the area be asked to look around their area and report unattended or out-of-place items? Before they can be asked to conduct a search, staff will need to be trained in basic search and awareness skills[31].
- Actually finding an item will assist both in confirming the validity of the threat and in giving the emergency services something to which to respond.

31   See Chapter 8, Search.

Other aspects that will be reviewed by the TET will include:

- possible motives behind the threat;
- the wording of the threat as recorded by the witness coupled with their recollections and impressions; and
- how much knowledge of the site the perpetrator displays.

If the perpetrator has actually made and delivered a bomb and now wishes to warn the organization it may be reasonable to assume that they will display knowledge of the event and the site.

It is important that the deliberations of the TET be recorded to assist with subsequent external and internal inquiries and investigations, to justify the actions taken at the time, and for the basis of future training and reviews of the process.

Respond

The questions to be answered are: "Do we evacuate or not, and how much of the site should be evacuated, and to where?". Protection from bombs is through distance and by sheltering behind something solid. If there is any reason to believe that the person may have placed a bomb as they claim in their threat, then an evacuation of all or part of the site should be initiated under the control of the site's Emergency Control Organization and in accordance with the Emergency Plan. See Chapter 9 for additional considerations in relation to evacuation and emergency procedures.

Evacuation plans should include the closure of processes as well as the emergency transfer of data and work processes in accordance with the Business Continuity Plan (BCP)[32].

The responses include:
- If the threat is not feasible:
    - Continue working,
    - Advise the police that a threat has been received and assist with any investigations.
    - Advise those staff and others that are aware of the threat that it has been carefully evaluated and the decision has been made to continue operations.
    - Complete the record of the evaluation and ensure it is filed for future reference.
- If the threat is plausible:
    - Advise the police and if necessary request assistance with any evacuation.
    - Coordinate with the Emergency Manager the evacuation of the site. Decide with the Emergency Manager the most appropriate evacuation routes and assembly areas based on the threat[33].

32   See Chapter 9, Emergency Management for considerations relating to BCP

33   See Chapter 9, Emergency Management

- Inform the responding emergency services of the location of the emergency rendezvous point.
- Consider security of the site during the evacuation.
- Consider when and how reoccupation of the site will occur, particularly if there is no explosion or other event.
- Depending on the threat, consider implementing the BCP. It is possible that the evacuation may last for some hours while the deadline passes, an appropriate safety period is applied and the reoccupation process commences, which may include a search of the site prior to staff re-occupying it.
- Complete the record of the evaluation and ensure it is filed for future reference.

Consideration should be given as to when the other tenants on the site should be notified.

The TET should have a prepared list of senior managers and others that should be promptly informed that a threat has been received and is being evaluated. The contact list could include the following, if they are not part of the TET:

- The head or regional office,
- The Site Manager,
- The site's Emergency Manager/Chief Warden,
- Facility Manager,
- The HR manager,
- The senior union representative, and
- The manager of the threatened site if operated by a contractor or service provider.

After hours or alternate contact details should be available for each of the nominated positions. It should be clear to those contacted that the responsibility for evaluating and responding to the threat rests with the TET. Having those on the contact list observe or participate in threat evaluation and response training will assist in gaining their understanding and support.

The response options are directly related to the nature and content of the threat. A full range of reasonable response options for the site should be identified, prepared and rehearsed. Consideration should be given to developing prepared media statements for release when threats are received[34].

---

34 See Chapter 9, Emergency Management for considerations on media management

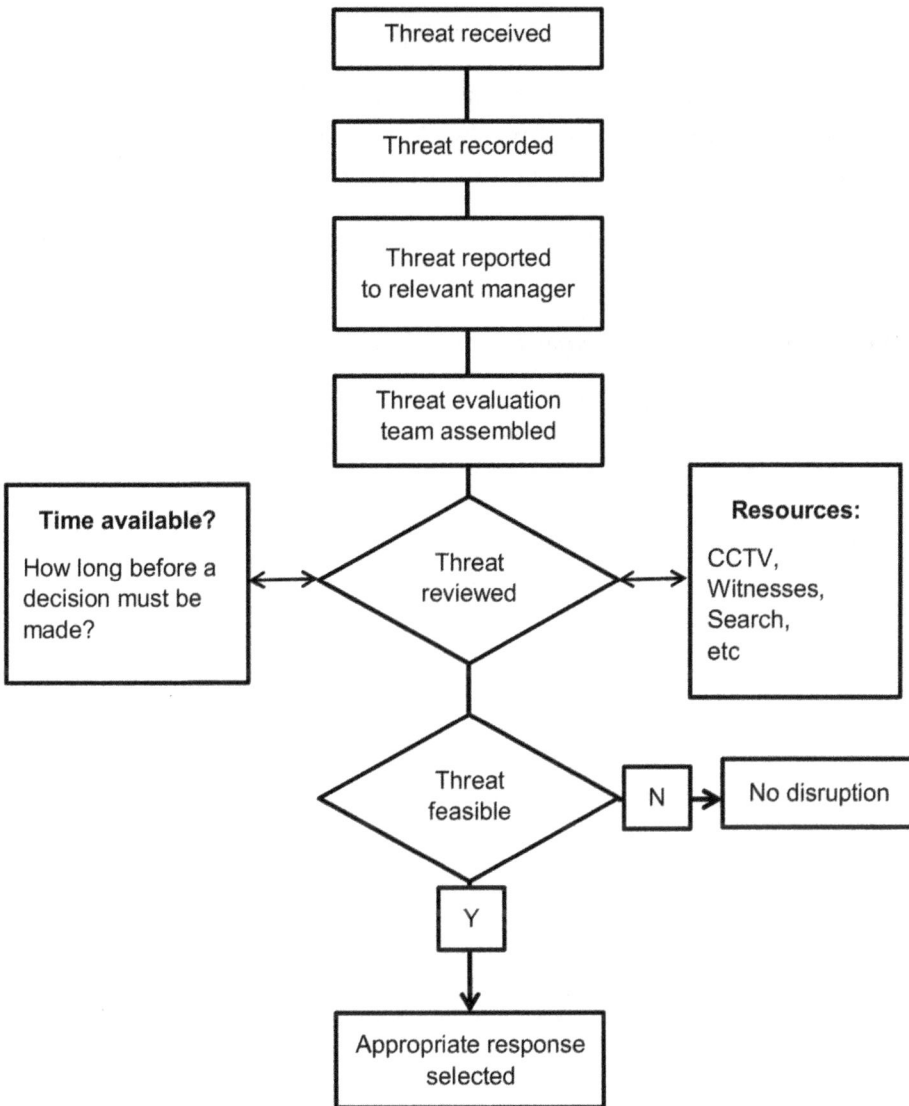

Figure 4 – Bomb Threat flow chart

Summary

Staff, contractors and possibly even visitors, should know that threats are treated seriously by management and that there is a process in place to evaluate them.

The key question when assessing threats is: "Can the perpetrator have done that which they claim?" This is a question managers should be able to answer with a degree of confidence.

All staff should know how to: **Receive, Record** and **Report** threats.

Managers, specifically the emergency control organization or equivalent, should know how to **Review** and **Respond** to threats. Those responsible for assessing the threats should know what resources are available to them, including the ability to have selected areas searched and also what response measures are appropriate.

Having an effective and efficient capability to evaluate threats of all kinds will reduce the likelihood of unnecessary disruption while increasing the potential to apply the appropriate safety measures should it be assessed that the threat is feasible.

Managers always have the option of evacuating if they believe this is the safest and most appropriate option.

Real world examples of threat assessments are provided in Annex C.

Checklists

The following generic checklists are provided to help organizations develop their own site-specific procedures and immediate response guides.

Planning Considerations

Bomb Threat - Planning	Response
A manager, and deputy, has documented responsibility for assessing and responding to bomb threats.	
Members of the team responsible for assessing threats, the TET, have been appointed.	
Bomb threat procedures have been written, approved and validated.	
Staff know how to Receive, Record and Report threats.	
The threat evaluation team has been trained and practiced in assessing bomb threats i.e. to Review and select the appropriate Response.	
Processes in place to record decision making and final decision.	

Incident Checklist

Bomb Threat - Action	Response
Threat received and recorded by whom and when.	
Threat reported to whom.	
Responsible person or team notified when.	
Deadline given in threat?	
Time available calculation possible – latest time a decision must be made?	
Is it feasible the person can do what they claim in the threat?	

Bomb Threat - Action	Response
Gather information from other sources?	
Deploy search teams Yes/No?	
Evacuate Yes/No?	
Document decision and reasoning behind it .	

Chapter 5

Preparing for and Responding to Unattended Items

A common occurrence that can lead to significant disruption is the 'unattended item'. This is an item that appears to have been abandoned. It may be a bomb or it may be an article forgotten, lost or temporarily placed there by the owner or even a piece of rubbish. Such items are more likely in public areas such as foyers, entranceways and outdoor areas. In most cases there will be doubt as to the nature and origin of the item.

Unattended items are also referred to as 'unidentified', which has the meaning of an item that cannot be recognised as an item appropriate to the site. The term 'unattended' is broader in scope and is used throughout this book to encompass items worthy of investigation and assessment.

Recognising, reporting and evaluating unattended items to decide if the site and occupants are at risk can be done quickly and discretely if the appropriate processes and training are in place.

Terminology is important; as well as 'unattended or unidentified' such items are sometimes referred to as 'suspicious'. To refer to an item as 'suspicious' before it is assessed prejudices any evaluation process. The word 'suspicious" has gained the overtones and implications of 'dangerous' therefore generating an imperative towards evacuation. The reality is that the item should be considered 'unattended' until it is assessed and determined to be 'safe' or 'potentially hazardous'. Potentially hazardous items are defined as containing what appears to be the components of a bomb or some other risk to safety and which is worthy of further investigation.

The likelihood of detecting unattended items increases if the staff are trained, aware and willing to report such items to supervisors/managers and when supervisors/managers know how to action such a report. Staff monitoring CCTV systems may also identify and report such items. The public of countries that have suffered from sustained bombing campaigns are quick to report unattended items.

This Chapter presents an unidentified item process (UIP) which includes the VALID methodology for the assessment of the item.

Unattended Item Principles

The question to be asked is: 'does this item pose a hazard?' A manager erring on the side of caution may decide to evacuate the site every time an item is found. If this were to happen there would be an impact on organizational productivity, staff, executives and customer confidence, the local emergency services, and the reputations of the organization and individual managers. Therefore a process is required to help the manager determine if the item may be hazardous.

What is considered out of place will vary from one area to another. An unattended school backpack in a public foyer, while of interest and worthy of investigation, may not be of great concern compared to the same backpack strapped to a gas cylinder. In the second example the evaluation process is quite simple and it is probable that a backpack strapped to a gas cylinder will be assessed as posing a hazard and an evacuation initiated.

The example of a backpack in a public foyer is more complex and requires a greater degree of evaluation. Video records may be accessed to determine how it got there, and staff and visitors may be questioned as to its origins. If it was a school bag, teachers escorting any visiting school groups could be approached.

Failing to identify a hazard places people, assets and operations at risk with subsequent reputational damage. Inappropriate response to non-hazardous items causes disruption to the organization and creates other risks as people are unnecessarily evacuated, emergency services are called when not actually required, industrial and business processes are closed, clients are inconvenienced, business continuity plans are initiated, etc. There is nothing safe about moving hundreds or thousands of people in a way they are not used to, unless they are being moved away from a hazard.

Advice is often given that an item should never be touched. If there is any reason to believe the item is hazardous then this is sound advice. But, if the item can not be seen into and there is no readily accessible portable X-ray system and if the item is more likely to be lost property or rubbish, then gently opening the item without moving it will provide more information such as yes, "the contents are typical tourist's belongings"; or, in the worst case, "this appears to be a bomb"; or "I don't know what this is but it does not fit my environment".

An unattended item can only be one of three things:
* Rubbish,
* Abandoned property[35], or
* Hazardous,

It is not the role of management or staff to decide that the item is a hoax - if it looks like a bomb or cannot be positively identified as safe then it should be treated as hazardous.

If the item is assessed as posing a hazard, the authority to initiate an evacuation under these conditions should remain with the site's emergency control organization (ECO) and the Emergency Manager[36].

Once the evacuation is completed the witness who saw the item should be available to brief

---

35   Abandoned property may have been: lost, left deliberately or temporarily unaccompanied

36   Note: the relationship between the ECO and security management and related disciplines (facility management, environmental management, safety, HR, legal, media relations etc) should be aligned as described in Chapter 9, Emergency Management.

the emergency services so they can deal with the item as quickly as possible, minimising disruption to the organization.

Some sites, assessed as being at high risk e.g. international airports, some government offices etc. have portable X-ray systems available to assist with investigating unattended items. For most organizations this will not be a cost effective capability.

The Five Rs

As with threats the 5R model may be used to address unattended items. The Threat Coordinator and TET can be trained to assess unattended items[37].

Receive

An unattended item is "received" when it is found by a member of the public, a visitor or staff member. All those on site should be encouraged to report anything they believe is out of place or unusual. Obviously their ability to do this will be related to their familiarity with the site and its operations.

Record

The person finding the item should record as much information as possible in as short a time as possible. Information of relevance is where is it, what is it, what does it look like, were there any smells, wires, aerials etc., as well as any activity in the area prior to the discovery of the item.

Report

The person finding an unattended item should know how to report it to the appropriate person, possibly a Fire Warden, Security Officer or supervisor. They, in turn should know how to report the item to the Emergency Manager.

Review

The review of the item to decide if it is safe or may pose a hazard needs to be done as quickly as possible. The VALID methodology explained below provides a structured means of assessing the specific unattended items.

Determining the circumstances of how the item came to be there will provide guidance as to whether it should be considered hazardous. In some cases it will be obvious from witnesses or CCTV footage that the item was accidentally left behind or discarded and should be treated as lost property or rubbish, alternatively it may have been deliberately secreted in a suspicious manner.

37　See Chapter 4, Bomb Threats

Respond

If the item is believed to be safe then it should be treated as lost property or rubbish, the final check being to look into the item to confirm its contents. If the contents indicate that it is hazardous, then the assessment is revised and an immediate evacuation of the area should be initiated in accordance with the site's Bomb Incident Management Plan.

The Unattended Item Process

The unattended Item process (UIP) provides an overarching structured system for identifying, assessing and responding to items that may pose a hazard. The assessment of the actual item is conducted using the VALID methodology as described later in this Chapter. The Process is in five phases as shown in Table 1 and Figure 5.

Phase	Description	Responsible	Comment
1	Identification and report	Public, staff member, contractor, tenant	If the item is identified by police they have responsibility for the assessment, go to Phase 3.
2	Reporting to relevant manager and tasking of an assessor.	Relevant manager	In cases where a trained assessor is made aware of the unattended item at the same time or before the relevant manager the assessor may begin the assessment pending tasking.
3	Assessment	Assessor	Use of the "VALID" process.
4	Response	Relevant manager. ECO if considered hazardous.	The appropriate response will depend on the whether a hazard is probable and the nature of the hazard. It is possible that resolution of the incident may require Emergency Control Organization and/or Crisis Management Team involvement.

Table 1 – UIP Phases and Responsibilities

Phase 1. Identify and report.

An unattended item may be identified by a member of the public, staff, contractor or tenant.

Members of the public may report the item to any person they believe to be associated with the organization/site.

Staff, contractors and tenants should know to report the item to the relevant manager.

If the item is identified by police, or reported directly to them, they may assume responsibility for the assessment and go to Phase 3. Police should advise the relevant manager of the unattended item and their assessment and response.

Phase 2. Reporting to relevant manager and tasking of assessor.

Once the relevant manager is advised of the unattended item they will task a trained assessor to conduct an assessment of the item.

The relevant manager will open an incident report and maintain a log of activities in accordance with site procedures.

In cases where a trained assessor is made aware of the item at the same time or before the relevant manager, the assessor may begin the assessment after ensuring the relevant manager is aware of the incident.

Phase 3. Assessment VALID.

VALID is a methodology used to assess unattended items. The mnemonic VALID is drawn from Visual, Assess, Labels, Information, Decide.

The use of a structured assessment tool provides managers and staff with a consistent and practical method of assessing if an item poses a hazard or, as is more likely, the item is safe.

VALID differs from previous processes in that:
* it recognises that most assessments are made quickly and efficiently, and
* it supports managers who decide not to evacuate because an unattended item has been reported.

The VALID Mnemonic

V = Visual inspection of the item and surrounding area.
What is it, where is it?

A = Assess: can a decision be made now? Is it:

Rubbish = dispose of it.

Lost or abandoned = treat as lost property.

Potentially hazardous = start evacuation and contact emergency manager.

If further information is required:

L = Labels, tags and other markings to identify where it came from and who owns it.

I = Information from CCTV, witnesses, others in the area, access control records, etc.

D = Decide, is it:

Rubbish = dispose of it.

Lost or abandoned = treat as lost property.

Potentially hazardous = start evacuation and contact emergency manager.

The Assessor's Actions

Specifically, the VALID actions for the assessor are:

V. Visual; an initial visual assessment of the item and its location. Is it obviously an item left by accident i.e. left on shopping trolley in a carpark, abandoned excess carry bag or rubbish? Does it obviously look hazardous: components of a bomb, white powder, plant or animal parts, military or commercial explosives, sharp items? Does the location suggest it may be hazardous i.e. strapped to a gas cylinder, placed under a car, concealed? Any information that is available at this point should be used to assess the item. The assessor should take a photograph of the item to assist in the assessment and for the record.

A. Assess; it may be possible from the initial visual inspection to determine if the item is obviously safe and does not pose a hazard or if it obviously suggests that it is hazardous. If a decision is possible at this point then the relevant manager is advised and they initiate a response in accordance with the appropriate emergency procedure. If the item is non-hazardous then it is: treated as lost property or disposed of as rubbish. If a decision cannot be made at this time then additional information may be obtained as follows.

L. Labels; are there any labels, tags, tickets or other markings that may help identify the owner? Nametags with phone numbers can be used to contact the owner. Labels and tags may help identify how the item arrived. It may be possible to determine if the item was purchased at a local retail outlet and, if so, information can be obtained from them.

I. Information; Additional information can be gained from a number of sources. Local witnesses, particularly employees, as well as nearby tenants and contractors can be asked if they saw the item arrive. Asking any visiting groups if they own the item;

using the site intercom to ask anyone who may have left an item to return to that area; checking access control records; and reviewing CCTV footage of the immediate and surrounding areas to identify how the item came to be there are other sources[38]. The security measures at the location where it is found will also be an indicator, as there may be a limited number of people who can have left or placed the item there. Others in the area may be able to identify the item as something that is common in their working environment and therefore not considered hazardous. If considered necessary the police can be requested to attend in which case responsibility for determining if the item is hazardous is transferred to them.

D. Decide; the end point of the assessment is a decision that the item is believed to be hazardous or non-hazardous. If the item is considered hazardous then request the relevant manager raise the ECO and initiate a response. This may require implementation of the Business Continuity Plan, etc. If assessed as non-hazardous, then the item is: treated as lost property or disposed of as rubbish. Ensure all reports are competed.

Most unattended items are not hazardous. In many cases the initial visual inspection of the item and its location will provide enough information to indicate the item does not pose a hazard.

Indications of a hazard include but are not limited to: wires, smells, unrecognisable contents, powders, plant, animal or other biological material, sharp objects, weapons, pyrotechnics, military or commercial explosives or ammunition.

The location can be an obvious indicator of a hazard such as the item being placed under or attached to a gas storage cylinder. In other cases the location may suggest that additional investigation is required such as a review of CCTV footage of the item and approaches to see how it came to be there.

Tools such as HOT[39], HOT UP, HOT ALERT may be used during the initial assessment phase if desired.

The assessor may request police attend to conduct additional assessment using tools such as X-rays and explosive detectors. Once the police are requested responsibility for assessing the item passes to the police.

Being aware of bombing trends through liaison with local law enforcement and through security networks may also assist in deciding if an item should be considered suspicious, particularly if it is similar to items reported elsewhere.

---

38   In some situations, portable X-ray machines may be available to enable trained and licenced staff to view the contents of an item

39   HOT = "Hidden. Obviously hazardous, Typical of the environment or similar phrasing.

Phase 4. Response

The assessor decides if the item is believed to pose a hazard or not. If considered non-hazardous, the assessor decides if the item is to be discarded as rubbish or treated as lost property.

If the item is considered hazardous then the ECO is raised and takes responsibility for resolving the incident in accordance with the relevant site emergency plan.

The ECO seeks assistance from appropriate emergency services.

It is possible that resolution of the incident may require involvement of the site's Crisis Management Team, Business Continuity or Resilience Plans, etc.

At the completion of the incident the report is completed. Detailed, accurate, contemporary reports will provide a sound basis for future analysis, improvement of processes, identifying training needs and, if necessary, the basis for explaining what decisions were made and why.

Unidentified Item Process

Phase 1

Unattended Item reported

By police

By public, tenant, contractor or staff member

Phase 2

Inform relevant manager

Relevant manager tasks Assessor

Initiate report

Phase 3

Assessment

Use VALID

Considered hazardous?

Y

Confirm type of hazard

Require more data?

N

CCTV, interviews, other sources

IED, HAZMAT, Chem/Bio, drugs, sharps, other?

Phase 4

Lost property

Rubbish

ECO initiate appropriate response

Complete reports

Involve Crisis Mgmt Team, BCP, etc?

Emergency Services take control

Figure 5 – Unattended Item Process flow chart

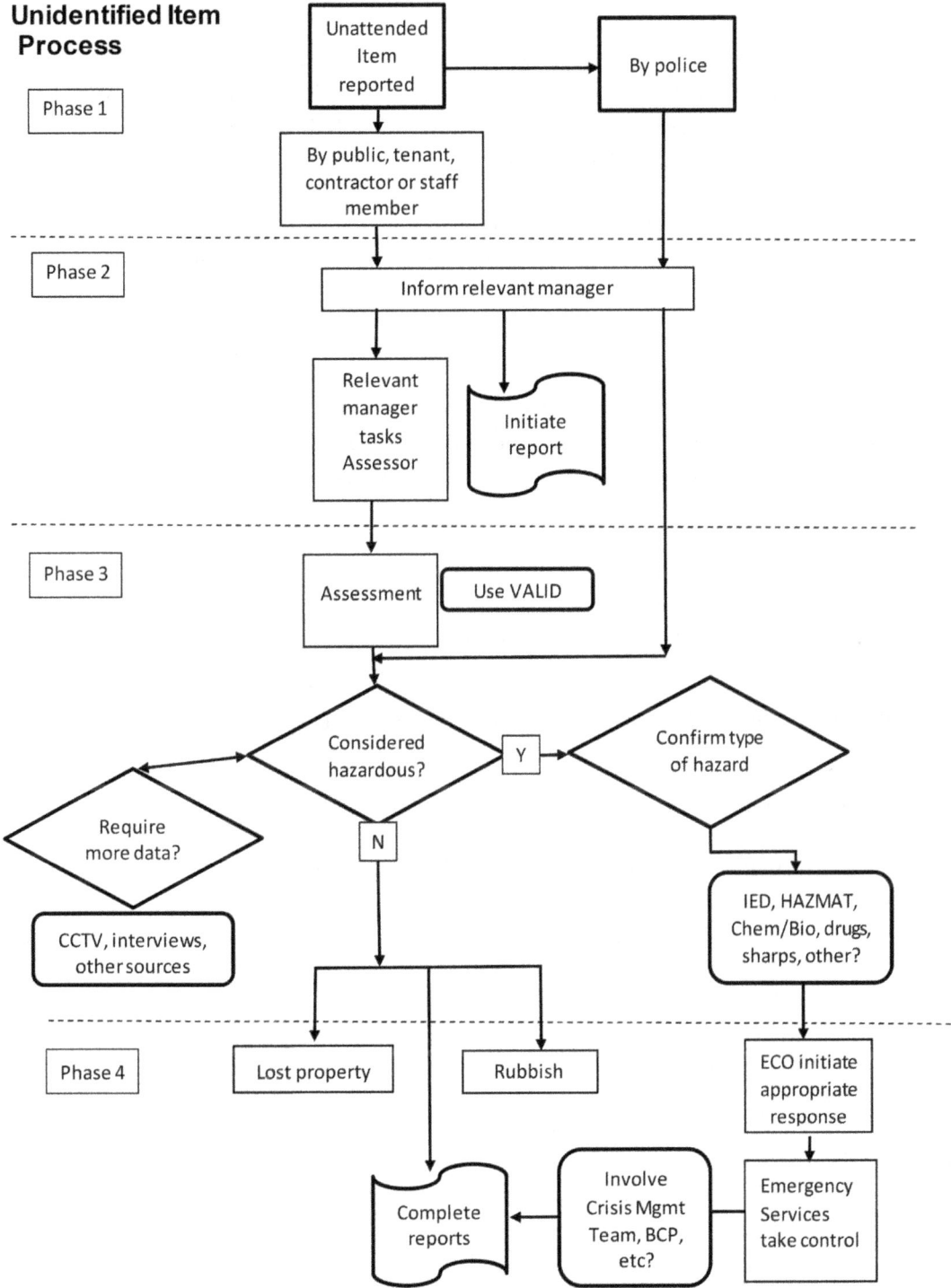

Summary

Unattended items are the most likely of bomb related incidents. They have the potential to cause considerable disruption if not assessed effectively and efficiently. The concern is that the item may be hazardous, specifically a placed bomb. It is more likely that it will be either rubbish or abandoned / lost property.

Evacuating when not required results in disruption to operations and can generate additional health and safety issues as people are relocated. Evacuation for a hazardous item can be expected to cause at least a half-day of disruption and loss of productivity.

Failing to evacuate when a hazard is present may result in death and injury.

Using the 5Rs, all staff and tenants need to know how to identify (Receive), Record and Report an item that they believe is out of place. Selected managers should be trained in how to Review the item to determine if it is rubbish, abandoned or poses a hazard, and to apply the appropriate Response.

Use of the VALID methodology will assist in reviewing the item: a Visual inspection of the item and surrounds, Assessing if it is safe or potentially hazardous, if additional assessment is required, checking Labels and other markings, gathering other Information about the item and how it came to be there and Deciding what the item is (rubbish, abandoned or hazardous) and the selecting appropriate response.

Managers always have the option of evacuating if they believe this is the safest and most appropriate option.

Real world examples of unattended item assessments are provided in Annex C.

Checklists

The following generic checklists are provided to help organizations develop their own site-specific procedures and immediate response guides.

Planning Considerations

Unattended Item - Planning	Response
A manager, and deputy, has documented responsibility for responding to bomb incidents.	
The person or team responsible for assessing unattended items has been appointed.	
Unattended items assessment procedures have been written, approved and validated.	
Staff know how to identify/Receive, Record and Report unattended items.	
The person or team has been trained and practiced in assessing unattended items i.e. to Review and select the appropriate Response.	
Processes in place to record decision-making, reasoning and final decision.	

Incident Checklist

Unattended Item - Action	Response
Unattended item identified/received and recorded by whom and when.	
Unattended items reported to whom.	
Responsible person or team notified when.	
Does the item appear to pose a hazard – use VALID methodology to assess the item.	
Gather information from other sources?	
Evacuate Yes/No?	

Unattended Item - Action	Response
Document decision and reasoning behind it.	

Chapter 6

Preparing for and Responding to Post-blast

If a bomb is not detected and disarmed by the emergency services, despite the best plans and measures, and there is an explosion, a 'post blast' response will be required.

The physical effects of a bomb can be similar to those of an industrial accident or natural disaster but, as a bombing is the result of deliberate human action, it is a crime and one that can have significant psychological, emotional and social impacts.

Post-blast actions will be aligned to the organization's Emergency, Crisis Management, Media, HR, and Business Continuity/Resilience plans. There are immediate and long-term considerations. An explosion will be traumatic, confusing, demanding and challenging. In some cases it is only after the investigation that will it be confirmed whether the explosion was a deliberate criminal act or an accident. The immediate responses will be similar regardless of the cause or motive.

It may be that the bomb was not targeted at the organization and that the organization suffered 'collateral damage'.

The effects of the explosion, the dust, darkness, falling ceiling tiles, broken water lines or if outside the scattered debris, smells, fires and surrounding damage will make it difficult to determine exactly what has happened.

The cries of the injured and those nearby will make it difficult to concentrate on determining the correct response. Managers may find that they are closely involved and it is their friends and co-workers who are injured and therefore find it hard to disassociate themselves enough to provide objective guidance and leadership. Prior planning and preparations will help.

Immediate Effects

In the immediate aftermath of a bomb explosion:
- The blast wave and fragmentation will cause injuries and possibly death to those in the immediate vicinity.
- Items near the explosion will be damaged resulting in projected fragments.
- Ceiling tiles will be dislodged adding to the damage and making movement and observation difficult.
- Glass partitions and windows may have shattered creating glass fragments.
- There will probably be smoke.
- Fire sprinklers may be damaged adding water to the scene.
- Cabling, piping and junctions may be damaged in the explosion causing problems

with power, fire-fighting systems, sewerage, communications and IT.

- Live electrical wires may be exposed and gas lines severed.
- First aid and immediate safety responses will be required as for an industrial accident.
- Poisonous and other hazardous substances may be generated by the explosive material or the building debris
- It is possible the building fabric may be weakened[40]. The site will probably have to be inspected to ensure the structure is still sound.

Immediate Response

The first few minutes after an explosion will be chaotic and there will be confusion, restricted communication with conflicting reports and a lack of clarity as to what happened. Prior planning will assist in Managers taking control of the situation and reducing the chaos.

The effects of an explosion can range from no casualties and little damage to numerous fatalities with a commensurately larger number of casualties and significant structural damage. The damage will depend on the size and location of the bomb and the intent of the bomber.

Until emergency services arrive, the priority must be on protecting those not involved and in providing first aid to the injured. One of the considerations for an explosion is that blast overpressure can cause injuries that are not immediately apparent such as brain, lung and other soft tissue damage. It is important to record the details of all involved and to ensure that everyone who felt the blast is inspected by the responding emergency medical service. The emergency services may also require details for follow up and investigative purposes[41].

People should be moved away from the site and protected from any falling debris, fires, structural damage or other hazards. Ideally people should be directed to an area where they can be segregated from the general public so they can be treated and their details recorded.

Where possible, bodies of the deceased should be left where they are as their position can provide vital evidence to the investigators. Also, each fragmentation injury and deceased body is a source of forensic evidence that may lead to the perpetrator.

If the incident is external to the site, the organization should have a 'Shelter in Place' capability as part of the Emergency Plan so that people can remain where they are until the nature of the hazard is identified and the most appropriate egress route and time for an evacuation can be determined.

40 See Yates 2012.

41 See Chapters 9, Emergency Management and 10, Training

Protection of the crime scene will assist in identifying the perpetrators. People should be kept away from the scene of the explosion. Bystanders should be pushed back as far as possible and preferably out of sight of the scene.

Due to social media it is probable that images of the scene, the damage and even of the casualties may be broadcast immediately. Staff should be advised, in prior training, that filming and sending images of any sort of accident or incident is inappropriate and could result in their being prosecuted, depending on the jurisdiction, or of having their camera confiscated by law enforcement as evidence. If possible, pre-arranged screens can be used to block site lines to the scene.

Medium-Term Response

Once the responding emergency services arrive, management has a role in supporting them and in assisting the staff.

- There will need to be a process to handover to emergency services and provide a briefing on what happened and what actions have been taken by the organization.

- Discuss with the emergency services what the criteria and process will be for handing some, or all, of the scene back to the organization.

- There will be a need for managers to maintain situational awareness of what is happening, what support is being provided, where the staff and others are being held etc. so that control of the incident can be maintained.

- The organization's Business Continuity (BCP)/Resilience plan may have to be implemented.

- There may be a need to have structural engineers and utility service providers such as electricity, gas, water, assess the site before it can be reoccupied or even before access can be granted to gather important business data, documents or other items. This should link to the BCP consideration of denial to the site. It would be of value for the managers to know how to contact appropriate engineers should their expertise be needed for a post-blast or other scenario[42].

- There will be a need to inform and update the owners, key stakeholders and in the case of government agencies the relevant head of Department. The corporate Media Plan will probably be initiated and the manager dealing with the post-blast situation will be expected to provide information to those fronting the media. The Media Plan should also address how it will manage the corporate implications of the explosion and immediate post-blast effects having already appeared on social media.

- Human resources (HR) will be required to provide immediate support to those who were on site. This may be assisted by the access control system records, if available. HR will also need to be ready to identify next of kin of those involved whether

---

42   See Gebbeken 2015

injured, killed or just on site. There will be a need for post-incident counselling in the near and long-term.

- There will be a need for post-traumatic incident counselling and support.
- The organization and the families should realise or be informed that the bodies of those killed may not be released until a complete examination has been completed.
- The organization's Media Plan should address how the medium-term implications of social media, including unauthorised releases, will be managed. See also Chapter 9 Emergency Management.

Long-Term Response

The impact of an explosion on an organization will be significant, and like other interruptions, the organization's response should be governed by its BCP. While most of the organization's responses will be the same regardless of the cause of the interruption, a bombing does introduce specific elements that may require additional consideration and response, primarily because a bombing is a deliberate act of violence. To account for these differences, the BCP should include a post-blast consequence management plan.

After an explosion, the site will be a complex crime scene. If there has been a fatality the investigators may isolate the immediate area for weeks while they sift through the evidence. There may be water and smoke damage to be repaired.

On-going medical monitoring of staff may be required. Secondary medical effects such as dust inhalation, trauma, and the severity of ear and eye injuries may not become apparent for some time. HR should ensure contact details of all involved are recorded as well as their involvement and the results of initial medical assessments. There may also be a requirement for HR support to staff for long-term health issues including monitoring for post incident stress.

In relation to the building, the availability of contractors, material and equipment to repair and refurbish the building and IT systems may be limited particularly if there were a number of buildings damaged.

Insurance payouts may be difficult to access depending on the wording of the policy, exclusions and the eventual determination if the bomber was acting for personal gain (criminally motivated) or a terrorist (politically motivated).

Legal should prepare for claims for compensation and possibly also of negligence, which may relate to the organization's bomb management planning, procedures, training and capabilities or lack thereof.

Legal should also gather and record as much information and evidence as possible as it may be some time before formal investigations are completed. It may be years before all the court cases and other inquiries are finalised.

Planning for Post-Blast

As part of the Bomb Incident management plan the post-blast consequence plan should address the following:

- Staff support functions, recognising that a bombing is a deliberate human act of violence and the psychological and societal effects on the staff and others may be considerable.

- If there are casualties the organization's Emergency Medical/First Aid and HR support plans will be required.

- Not all blast injuries are immediately apparent. Anyone exposed to the blast must be recorded as being present and be examined by qualified medical personnel.

- The scene of the explosion will be a complex crime scene and may be isolated by the investigating authorities for a considerable period of time.

- Urgent structural assessments of the building may be required. The plan should state who will do these and have required engineering companies identified, including after hours contact details.

- The ability to obtain repair services, parts and equipment maybe restricted, particularly if the explosion affected a number of buildings in the area.

- Insurance may not cover the bomb damage depending on the exclusion clauses in the policy, in which case other means of funding repairs and replacements may be needed.

- Legal protection from litigation or claims of negligence may be required. A demonstrated adherence to pre-planned bomb incident security management practices may assist with a legal defence.

Similar considerations to those above may also be used for planning consequence management for explosions due to industrial accident etc.

Summary

A post-blast situation will be confusing and challenging. The first priority should be protection of life, followed by protection of the crime scene and protection of the organization. In the longer term there will be a need to monitor and to provide support to those involved.

There will be immediate, medium and long-term responses required to a post-blast incident. The survival of the organization may depend on how these issues are managed. Considering the implications of an explosion, regardless of the cause, and planning for the aftermath will assist.

The post-blast response will rely on other corporate plans and post-blast considerations should be a factor when those plans are being drafted, reviewed and exercised.

Real world examples are provided in Annex C.

Checklists

The following generic checklists are provided to help organizations develop their own site-specific procedures and immediate response guides.

Planning Considerations

Post-Blast - Planning	Response
A manager, and deputy, has documented responsibility for responding to bomb incidents.	
Post-blast scenarios considered.	
Post-blast planning aligned to organization's BCP/Resilience plan.	
Insurance for bombing considered – terrorist action excluded? Options?	
Structural engineers, urgent repairs – identified and available.	

Incident Checklist

Period	Response	
Immediate	Move people and injured away from danger. Look for hazards from falling debris, structural damage. Administer immediate first aid. Move bystanders away. Record all those involved.	
Medium Term	Handover to emergency services. Possible implementation of BCP/Resilience plan. Structural inspection of site to allow urgent access. Media management.	

Period	Response	
Long Term	Structural inspection.	
	Repairs and replacement.	
	Insurance.	
	Media management.	
	On-going medical, mental and other staff support and rehabilitation.	
	Legal issues, claims, litigation, prosecution.	

Chapter 7

Preparing for and Responding to Hazardous Mail

Mail bombs are only part of the spectrum of hazardous mail (HAZMAIL) that could be received at workplaces. Hazardous mail includes; explosive and incendiary material, noxious and poisonous substances, acids, chemical or biological agents, radiological material and "sharps" such as needles or blades.

The primary protection against hazardous mail is recognising it before it is opened. The objective of an effective mail screening process is to identify those items which are different from those typically received. Hazardous Mail processes should inform staff and managers how to: Receive, Record, Report, Review and Respond to all types of hazardous mail.

There are occasions when hazardous items are sent through the mail without the intent to cause injury or fear; biological material, broken equipment or samples of soil or animal products may be sent as part of a complaint or application. It is legal to send some hazardous items through the mail system as long as they are correctly packaged and marked. These contents are still hazardous and need to be identified and managed.

The use of on-line purchasing has increased the number of packages being delivered through postal and courier systems. This adds to the items to be inspected at the workplace and also to the variation of items delivered and hence what is 'normal' for the environment. It is recommended that staff not have purchases delivered to their work place.

Types of Hazardous Mail

There are four types of Hazardous Mail:

- Mail bombs (including incendiaries).
- Chemical and biological.
- Sharps.
- Radiological.

Courier-delivered items are included in this Chapter as they raise additional considerations.

Each type of hazard has different indicators and requires different response measures.

Hazardous items sent through the mail are normally designed to function upon opening, whether it be exposing the sharp objects, releasing material or triggering a bomb. The injuries caused by sharps, poisons and noxious items will be limited to a relatively small population and may be easily contained. Mail bombs and items containing chemical or biological material will impact on a larger number of staff and have a greater impact on the operation of the organization.

HAZMAIL Procedures

The Five Rs

Hazardous Mail procedures should cover:

- Identification and reporting of the suspected hazardous mail item (**Receive**, **Record** and **Report**).
- Investigation as to why it was considered to be suspicious (**Review**).
- Determining if it is considered hazardous or safe to open (**Review**).
- If it is considered hazardous, identification of the type of hazard and the appropriate response measures (Review and **Respond**).

The critical element rests with the person who receives the item and is expected to open it, which is why all staff should receive training in hazardous mail identification and appropriate response procedures.

The families of senior executives, high-profile corporate members and other staff assessed as being at higher risk because of their roles and responsibilities should also be instructed in mail screening. There is a possibility that they may receive hazardous items at home.

Having as few sites as possible where mail is opened lessens the areas at risk and the number of staff requiring detailed mail-screening training. Ideally no mail should enter the work place without having been screened. Where possible, the screening point should be located where the discovery of an item believed to be hazardous will result in minimum disruption to the operation of the site.

Unless every item is opened during screening there is the possibility that a hazardous item may not be detected. Therefore all staff should receive training in the identification of, reporting and immediate response to hazardous mail.

Some organizations transfer the risk by having a contractor receive, inspect and open the mail. In such cases it is of value to ensure the service provider has appropriate response measures in place to minimise disruption to the mail delivery system.

Receive, Record and Reporting of Hazardous Mail

It is possible to visually identify many items of hazardous mail, as they will have characteristics different from normal mail. As with any type of bomb search, staff should be looking for that which is out of place or that which does not match the work environment. Millions of mail items are handled every day and staff are aware of what the mail received by the organization normally looks like. The objective of an effective mail screening process is to identify those items that are different. All mail and courier items should be inspected to determine if they pose a hazard. This inspection can usually be done quickly as most items will be 'normal' and fit the organization's operating environment.

'Mail bomb' identification guides are provided by government organizations and are based on historical examples. The indicators are based on the identification points expected for an explosive or incendiary device and form a sensible base for identifying other types of hazardous mail.

Bombs have a number of basic components: a power source, often a battery which can give the item an uneven balance; a firing switch, normally victim operated for mail delivered items, this switch may contain wires or foil; and an explosive or incendiary filling, which will add to the weight and may exude an oil. The whole device may be mounted on a piece of card or wood and be heavily taped or tied to stop it coming apart. The sender will try to ensure that the item is received and opened by the intended victim, hence the excessive postage, restrictive markings and distractions to prevent proper examination. People who send hazardous mail may not be in a balanced frame of mind, resulting in a history of misspelt and incorrectly addressed envelopes. Also, it is possible the sender has made and wrapped the bomb before addressing it, which may explain the poor handwriting. Hazardous mail containing sharps may have a rigid feel or bumps or uneven surfaces. Mail containing biological matter, acids or other liquids will not feel like normal documents.

The following identifying features may indicate the item contains some form of mechanism and/or hazardous material:

- Protruding wires or foil
- Oily stains
- Unbalanced
- Stiff or ridged packaging
- Soft or malleable contents
- Lumps that might indicate batteries
- Excessive wrapping'
- Excessive postage
- Incorrectly addressed
- Poor handwriting
- Any indication that the item is not the usual type of mail received by the organization

These are all valid indicators of items that may be out of place but in some environments some or all of these indicators may be normal. The Returns section of a production plant can expect broken parts with oily stains, protruding wires or excessive wrapping on a daily basis. At pathology laboratories biological material delivered through the mail may be normal, if it is packed appropriately.

None of the identifying features mean that the item is hazardous. What they indicate is that the item is worthy of further investigation to determine if it may pose a hazard or is safe to open.

The experience and opinion of the person inspecting the mail should always be valued. It may be that the person knows the item is unusual but cannot explain why. In such cases the item should be subjected to additional investigation. While identification guides can be of use they must be modified to reflect the operating environment of the organization.

All staff should be given a base level of awareness on how to identify unusual mail items and how to report them and to whom. Mailroom staff need a higher level of awareness as it is assumed all mail will pass through the mailroom.

Staff should know that if they have any concerns about an item of mail they should place it down and report it to the appropriate person as per the site's security and emergency procedures: security staff, local fire warden or manager. The staff member should be ready to show where the item is and to discuss their concerns.

The local security officer, warden or manager should know the next steps in the process, whether it is their responsibility to investigate the item or if they report the item to someone else.

The site-specific considerations will depend on the size of the organization, the amount of mail received, whether it is opened in a central mail area or by individuals at their desks, what screening capabilities exist and the security and emergency management structures within the building.

Review of Hazardous Mail

If the item is considered out of the ordinary it should be subjected to additional investigation.

There are a range of investigation methods that can be applied to determine if the item should be considered suspicious. Noting that if the item has arrived though the mail system it is unlikely to have a time switch, hence there is time to gather additional information and calmly assess the situation. Sources of information include asking other staff or supervisors if they recognise the item, and verifying the validity of any return address, particularly if a company name is provided. A key question is asking the recipient if they are expecting the item.

Each organization will be able to identify other sources of information to confirm the validity of a mail item. For a courier-delivered item, the courier company can be asked to check the origin of the item - this can be done before or after an evacuation, depending upon the level of concern (see Courier Delivered Items, below).

Equipment can assist in the screening of mail however the capability of the equipment varies, usually according to price, in both capital terms and training cost. Each system has advantages and disadvantages and must be tailored to the particular workplace. In all cases the equipment should be used to augment the skills of aware and trained staff.

Transparency sprays are available which can be applied to envelopes to provide a degree of visibility of the contents. These sprays are relatively inexpensive and can be effective on paper envelopes and wrapping. They do not work on all materials and can damage the documents inside. In some cases using a spray to wet the packaging could weaken the wrapping and therefore increase the hazard. They may be of benefit when a fast and low cost capability is required for occasional use.

Metal detectors are designed to indicate if there is any metal content within the item and come in a range of styles. Metal detectors can be hand-held or belt fed. The presence of metal does not mean a hazard exists as normal mail often contains metal in the form of clips, binders, computer components or packaging. There are discriminatory metal detectors which do not alarm at most stationery items, although bulldog clips, because of their looped wires, can still register. Hand-held metal detectors used at personnel screening points are not designed for mail screening and unless their limitations are known and allowed for they should not be used as they are likely to create a false sense of security. Metal detection is a first step in segregating mail that requires additional investigation; it does not confirm the presence or absence of a hazard.

There is a range of explosive detectors available. Most work by taking a vapour or physical sample and subjecting it to a chemical analysis. These systems are designed as investigative tools where the presence of explosives is suspected and confirmation is required. Explosive vapour detectors (EVD), which are based on analytical technology such as gas chromatography, are relatively expensive and require staff to be properly trained in their operation. Some have the advantage that they are dual technology and can also be used to scan for drugs, which may be of benefit to some workplaces. There are simplified systems such as explosive detection sprays, which indicate the presence of explosives by spraying the sample with a reactive indicator. The sprays are relatively inexpensive and provide a reasonable investigative capability. These systems are not intended for bulk screening and are best used as secondary screening systems to test items identified as suspect. In some cases a series of tests may be required for the different types of explosives. At this stage there are no explosive detection systems that can quickly and accurately screen large amounts of mail. While of benefit, explosive detection systems will not detect other forms of hazardous mail such as sharps or chemical and biological hazards.

The ability to look into the package without opening it enables the detection of most hazardous items. X-ray machines come in a range of sizes and types ranging from small portable items that can be linked to digital image capture screens to cabinet X-rays and belt-fed systems similar to those seen at airports. The volume of mail being processed will determine the size and capacity of the machine, for example a small cabinet machine may be of benefit to small or medium size enterprises. A mailbag can be quickly scanned in a cabinet and unidentified items isolated and subjected to additional screening. X-rays can also screen quite large objects such as cartons. Organizations may find additional uses for an X-ray machine, for example the ability to inspect some equipment or the non-intrusive screening of briefcases prior to sensitive board meetings. The X-ray machine does not need to be located in the mailroom, it may be of more benefit screening incoming deliveries or out-going items for inappropriate contents with the mail being taken to the machine once a day. There are health and safety implications relating to the use of X-rays, and organizations must ensure the X-ray machines meet all legal requirements and are maintained by qualified and certified agents. Effective use of X-ray machines does require staff to be trained in X-ray interpretation. If after X-raying, the item is believed to pose a hazard it should not be left in the X-ray machine where it is surrounded by large metal components and possibly a live X-ray source. If believed to contain a mail bomb or sharps, it can be moved to an isolation area (see below). If the item was courier-delivered or is believed to contain chemical/biological or radiological material then it should be moved out of the X-ray machine to the nearest flat surface.

The detection of biological or chemical material in the mail is more difficult. Many of the hazardous mail identifiers will still be valid such as excessive postage and securing material, personally addressed with misspelt titles or names. The item may feel as if it contains grit, powder, liquids or other 'unusual' contents. It is important that such items be detected prior to opening.

Business and government departments may receive unexpected biological or chemical material in support of a claim or complaint. Examples include bodily fluids and samples, including human faeces and blood, being submitted to support claims of medical complaints or sightings of animals. In other cases body matter has been sent to parliamentarians as a sign of disproval. Soil samples, believed to be contaminated by chemicals, have been sent through the mail. The receipt of such items will cause disruption. Those who have come in contact with the samples may be distressed and should undergo medical testing, which can add to the level of concern and the site may have to be cleaned. Processes to detect and respond to hazardous mail should be able to identify such items and reduce the risk of them impacting on the staff and operations.

The primary protection against hazardous mail is recognising it before it is opened. The best identification for chemical and biological hazards is trained staff, supported by effective screening equipment and response procedures.

Response to Hazardous Mail

The various hazards that can be sent through the mail have different identification and response measures. Hazardous mail tends to function when the envelope or packaging is opened and the hazard is exposed.

The organization is capable of assessing unusual mail items to determine if they may pose a hazard. If the site was to be evacuated and the emergency services called to investigate the item every time an unusual item was identified productivity would suffer, the emergency services may comment on the frequency of call outs and the capability of the relevant managers may be questioned. As mentioned in Chapter 9, Emergency Management, evacuation of hundreds or thousands of people is not always the safest or most appropriate response.

If there is no indication that the item is hazardous then it can be processed as normal mail.

If an item is believed to be hazardous there are a number of actions that should and should not occur. The following generic DOs and DON'Ts should be observed by all staff:

- Do NOT open it, even to see if it really is dangerous. It is the act of opening it or altering the packaging that will probably cause it to explode, to release its contents or to expose the sharp objects.
- Do NOT wet the item (including immersing it in water) as this is likely to alter the packaging and may cause the device to function or the dangerous material to be released.
- Do NOT place the item in a container, particularly not in a locked one. This makes it difficult for the responding emergency services to access it and adds to the fragmentation risk.
- Do NOT invite other people in to look at the item. Once the investigation reaches a stage where it is considered hazardous there is no need for additional people to be exposed to the potential risk.
- Do NOT carry the item through crowded areas. If it is to be moved choose a route that will cause minimum disruption to the rest of the work force or public.

The following actions should be undertaken for hazardous mail items:

- DO investigate the item as described in this Chapter.
- DO place the item on a flat surface away from similar items so it can be readily accessed by the responding emergency services.
- DO consider moving the item to a pre-designated isolation area as described in this Chapter.
- DO gather as much information as possible about the item and its location for the responding emergency services, including why it is considered hazardous.
- DO evacuate the immediate area using the existing workplace emergency procedures

(unless an isolation area is used).

- DO report it to the relevant local authority, usually the local area warden who will pass the information to the Emergency Manager. Alternative points of referral include the mailroom supervisor, the security manager, the OH&S officer, etc. The critical thing is that the manager(s) with the authority to control the incident are informed.

- DO report it to the emergency services, usually done by the Emergency Control Communications Officer or Emergency Manager. It should be noted that the police must be informed as an offence has been committed, probably including offences against the local Postal Act.

The above is generic guidance; each type of hazardous mail will have different identification features and additional response measures.

Specific Considerations for types of HAZMAIL

Bombs in the Mail

An item may be assessed as containing an explosive device if it seems to:
- contain components,
- be oily or feels like it has unusual filling,
- have unbalanced contents,
- have items mounted to a piece of cardboard or similar, or
- have any other feature or combination of indicators.

The primary step is to confirm the item has been delivered through the postal system. Items delivered through the postal system have been through a robust distribution process that demonstrates the item can withstand constant handling. As the exact time of delivery cannot be predicted, the sender is unlikely to have used a time-delay fuze. Similarly, the item is unlikely to function unless it is opened or the outer packaging is changed in some way. Therefore, the item is safe to move, so long as it is done carefully and without damaging the packaging. Also, there is time to conduct an investigation into the origin of the item.

If it cannot be confirmed that the item has been delivered via the postal system then the above assumptions are not valid. Consequently, the item should be treated as 'Courier-delivered', it must not be moved and an evacuation of the area should be immediately initiated.

Isolation Area

It may be possible to identify an isolation area where mail items can be placed while they are investigated. This enables work to continue with minimum disruption. At sites where

X-ray or other screening equipment is available the isolation area can be used after the investigation.

Should the item be considered hazardous the isolation area can be used to store the item until the emergency services arrive. As the item has been through the mail system it is considered safe unless acted upon e.g. opened, therefore there is no requirement to evacuate staff until advised to do so by the emergency services when they carry out additional investigative and intrusive measures.

An isolation area should be:

* away from the general work areas so as to cause minimum disruption should the item be assessed as being hazardous,
* not in a high pedestrian traffic area,
* not near any flammable or other hazardous substances,
* protected from access when used to isolate an item,
* protected from the weather,
* ideally under CCTV coverage so the item can be observed,
* provide easy access for any responding emergency services.

If an isolation area is not available, even if it is understood that the device will not function unless the envelope or package is opened, it may be appropriate to evacuate some of the site during the investigation to provide peace of mind to the occupants.

If an isolation area is not available then the item should be placed on a clear flat surface where it will be immediately obvious to the responding emergency services and the immediate area evacuated.

Chemical, Biological, 'White Powder'

'White Powder', or more accurately chemical or biological (CB) contamination of mail items, has become an area of increased concern after the live anthrax incidents in the USA in September and October 2001. It should be noted that the sending of powder or other material with notes claiming that the substance is a CB agent is not new. These intimidating incidents have been occurring for decades. The chances of receiving a hoax are considerably higher than being the recipient of a live CB item. In all cases, the incident must be handled effectively and efficiently.

Legitimate biological agents, permissible for carriage, are delivered to medical centres and laboratories every day. They constitute part of the normal traffic for those organizations and procedures to deal with them and any leaks or spillage should be part of the organization's emergency plans. The main concern is with biological or chemical material sent with the deliberate intent of creating harm or fear.

The effects of hazardous biological or chemical material introduced to the worksite via mail will be insidious and invasive. Once exposed the material will enter the buildings airflow and will contaminate the surrounding area including the skin and clothes of those nearby. Depending on the type of material the onset of the illness or chemical effects will range from instantaneous to long incubation periods. Staff, clients, contractors and other visitors to the site will be at risk. Apart from the need to treat those obviously affected, a wider population will need to be examined and probably treated and decontaminated. There may be far reaching medical implications involving National, State and local medical resources.

In most cases chemical or biological material will not be detected until after the item is opened and the material has been released. In some cases there may be additional indicators such as the feel of powder, fluids or capsules that suggest the presence of chemical, biological or similar material.

There are two aims when dealing with possible chemical or biological agents. The first is to limit the distribution of the material, the second aim is to look after the staff who have been exposed.

To prevent the spread of airborne material:
- Cover the item where it is. For example, large bags or bins can be placed carefully over the material, thereby trapping the contaminant with minimum disturbance.
- Design or position the mailroom so it can be isolated from the site's air conditioning or reticulation system.
- If mail is to be opened in the mailroom, consider the use of specifically designed fume and containment cabinets where mail can be opened and any contaminants will be trapped, ensuring they meet national standards for containment of biological and chemical hazards.
- Consider the use of containers and bags in which items suspected of being contaminated can be placed.

Person(s), having limited the spread of the material, should:
- Leave the contaminated area.
- As a precaution they should not join the general population rather they should be moved to a nearby room, preferably one that has access to wash facilities and hands-free phones.
- As the responding emergency services will wish to speak to these people and assess whether they need to be decontaminated, access to hands-free phones should be considered in the planning phase.
- Generally, staff should not brush or remove clothes, as this will disturb any material.

- Staff should be encouraged to wash their hands and trap the water for later analysis and disposal.
- Record the details of all who were on site at the time, (particularly those who may have been contaminated) for use by the emergency services[43].
- Counselling should be offered to all affected personnel.

Sharps

If sharps are identified in an item of mail:
- Place the item down.
- Render first aid if required.
- Notify the Emergency Services.
- Consider blood tests for possible infection.
- Counselling should be offered to all affected personnel.

Radiological

If a radiological hazard is identified, possibly through a number of legitimate items marked as containing radiological material accumulating in the mail room or other area:
- Place the item(s) down.
- Evacuate.
- Record the details of all who were on site at the time, particularly those who may have been contaminated for use by the emergency services.
- Notify the Emergency Services.

Counselling and staff support may necessitate monitoring and support for post-incident stress.

Courier-Delivered Items

Courier-delivered items offer additional challenges. They are often delivered wrapped in the courier company's packaging thereby disguising many of the identification features. Also, couriers will seek to deliver items within a specified time so there is an increased possibility of a timed fuze. If the courier is the bomber there may be an anti-handling switch armed as it is delivered and therefore the item should not be handled.

The same screening principle of looking for that which is out of place is applied for courier delivered items. Recipients will often know what types of packages are expected at what times. Consideration should be given to having:
- All couriered items delivered to one location where the staff may be familiar with the

---

43   A record of who was on site may also assist with later claims for compensation, claims of negligence, etc.

couriers and the types of deliveries expected.

- Staff confirm that the item is expected when collecting couriered items from the mailroom or similar.

- Mailroom personnel deliver the item after inspection and confirmation with the recipient, if internal organizational procedures do not permit staff to attend the mailroom.

- The courier's bag or envelope could opened and the inner contents inspected for identification features before being forwarded to the recipient(s).

Ideally and where cost effective, all courier-delivered items should be inspected by an X-ray machine.

It is useful to know if the courier staff place the items in the courier bags themselves. If so, the risk of opening the outer bag is greatly reduced and only the inner items need to be subjected to inspection. If the courier company accepts items already wrapped in their bags then the outer bag and label should be screened. Emphasis should be placed on factors such as the weight, shape and balance. Staff receiving courier-delivered items should also observe if the courier is acting suspiciously.

Items delivered by a courier offer a significant disadvantage to the sender, as there is often a greater chance that the item can be traced back to the place of origin. Even though the likelihood of receiving hazardous items via a courier may be less than through the mail, all staff must be made aware of the need to screen ALL delivered items.

If there is any indication that a courier-delivered item is hazardous, it should be treated as a bomb, not as a mail bomb, as it could contain a time fuze or other type of triggering switch. It should NOT be handled and the area should be evacuated immediately. The processes for dealing with a bomb should be implemented, the Emergency Control Organization contacted and the corporate emergency procedures instituted immediately.

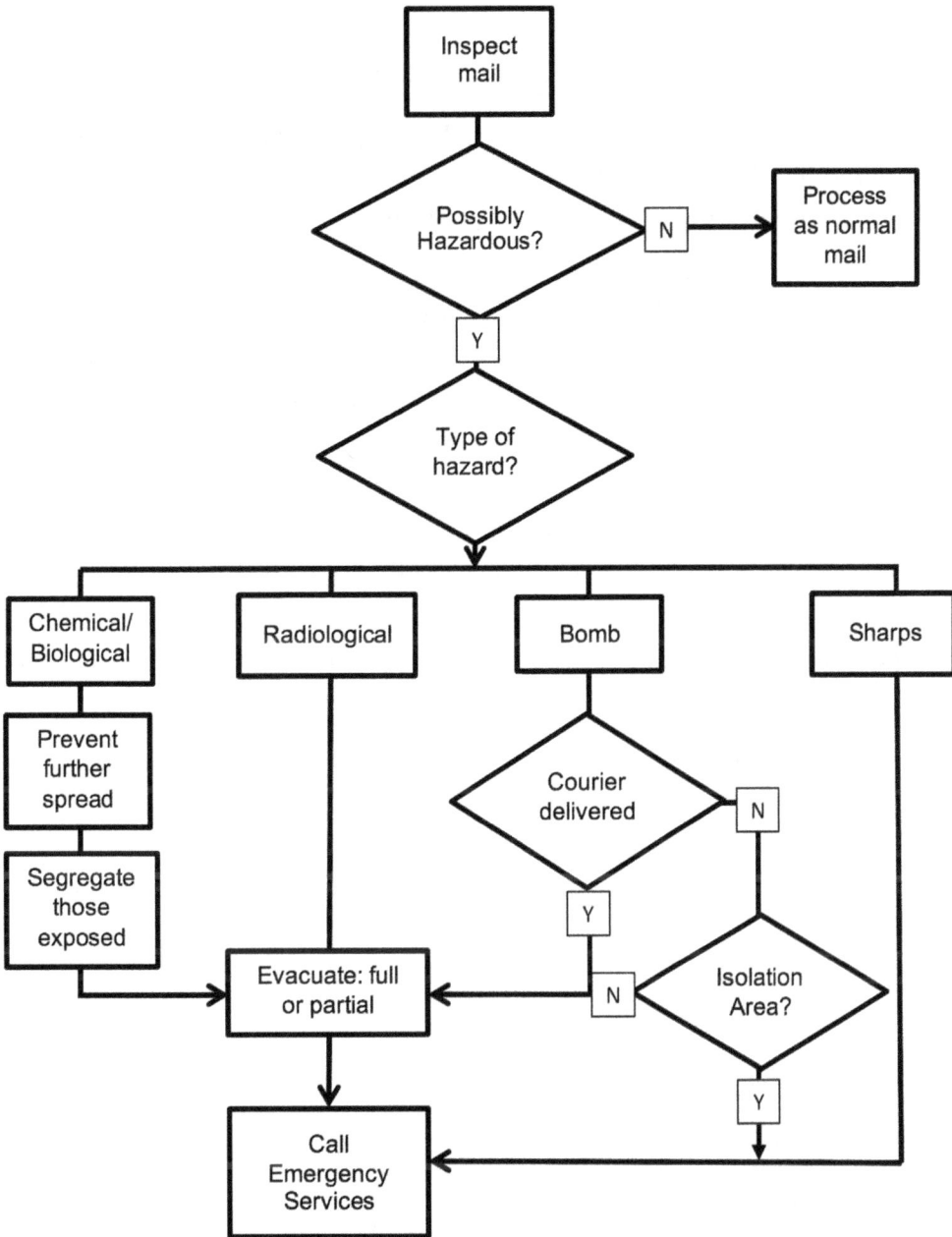

Figure 6 – Hazardous mail process flow chart

Summary

There are four types of Hazardous Mail:

- Mail bombs (including incendiaries).
- Chemical and biological.
- Sharps.

- Radiological.

Courier-delivered items raise additional considerations.

The primary protection against hazardous mail is recognising it before it is opened. The best identification for chemical and biological hazards is trained staff, supported by effective screening equipment and response procedures.

- For mail bombs – ensure the item has been through the postal system, place it on a clear flat surface, if possible use an isolation area, or evacuate the immediate area.
- For chemical/biological hazards, cover the material to prevent further spread, move those that may have been contaminated away from the item but do not allow them to mix with the general population, evacuate the rest of the area and record who was on site.
- For radiological, evacuate the area and record who was on site.
- For sharps, apply first aid if necessary, place it on a clear flat surface, and use an isolation area if possible.
- For courier-delivered items respond as for a placed bomb rather than a mail bomb as it may be timed.

Managers always have the option of evacuating if they believe this is the safest and most appropriate option.

Checklists

The following generic checklists are provided to help organizations develop their own site-specific procedures and immediate response guides.

Planning Considerations

Hazardous Mail - Planning	Response
A manager, and deputy, has documented responsibility for responding to bomb incidents.	
HAZMAIL procedures have been written approved and implemented.	
All staff are trained in identifying, reporting and responding to the different types of HAZMAIL.	
Mailroom, mail-handling staff receive additional training.	
Isolation area identified and documented.	
Ability to isolate chemical biological material anywhere in the work place Yes/No?	
Separate mail screening and handling room Yes/No?	
Ability to isolate air reticulation system in the mail room Yes/No?	
Liaised with local emergency services to understand their requirements and expectations and to explain organizational responses.	
Processes for briefing and handing over to responding emergency services.	

Incident Checklist

Hazardous Mail - Action	Response
Hazardous mail item identified.	
Investigate to determine if hazard may exist.	

Hazardous Mail - Action	Response
Determine type of hazard.	
Check if it has been through postal system if, not apply courier-delivered response.	
Apply appropriate response depending on type of hazard: mail bomb, courier delivered, sharps, chemical/biological or radiological.	
Advise emergency services.	
Meet emergency services at emergency rendezvous point (see Chapter 9)	
Hand back from emergency services after incident finished	
Implement safe reoccupation process	
Internal and external reports completed and submitted.	

Chapter 8

Search

Using staff to search areas is a valuable tool when assessing a threat. But a distinction should be made between staff looking around the area and those trained in search. The difference relates to the degree of certainty that the area has been properly inspected.

'Search' is a skill that can be taught to staff and which offers a much higher degree of confidence that an item will be found or that despite a dedicated, coordinated and comprehensive search of the nominated area, nothing out of place was identified. When, upon leaving a room or area, a search team reports that "nothing was found" the search coordinator must have confidence that the area was properly inspected. Staff who search should be volunteers, as it can be perceived as a risky activity. The provision of training can overcome concerns if it explains what is expected, how the search will be managed, that their safety is paramount and that the search will provide valuable information that will assist in providing a safer and more efficient work place.

The key benefit of a search is that finding a bomb provides confirmation that the evacuation is warranted and provides valuable information to the responding emergency services so they can deal with the bomb, hopefully before it explodes. Unless a possible bomb is located there is no basis for the bomb squad to attend.

Searches may be conducted before, during or after an evacuation. The decision on when to evacuate and when to search should be made by the Emergency Manager and will depend on the threat evaluation and the type and business of the site, the time of day, the current threat environment and related considerations.

Why Search?

Search should be part of a considered response. The reason why a search is being conducted must be clearly stated. Usually a search is used to gain additional information[44]. In relation to a threat search may be used to determine if an item has been placed or if there are signs of forced entry or tampering with doors, windows, storage areas or containers for example.

If there is a reasonable belief that there may be a bomb on the site, and there is adequate time prior to the deadline[45], some time may be spent trying to find it so the bomb squad has something to respond to.

If the site is evacuated due to a threat and a bomb has not been found, the option is for the site to remain evacuated until after the deadline, if one was given, plus a safety margin and then to reoccupy the building. If no deadline was given and no bomb was found, the

---

44   See Chapter 4, Bomb Threats

45   See Time Calculation in Chapter 4.

Emergency Manager will need to decide when the site can be entered. If an evacuation was called and the given deadline passes without incident, managers may consider, after a suitably safe period, asking the search teams to search all or part of the site before reoccupation.

Prior to a special event, such as the visit of a VIP, a search may be used to ensure there is nothing out of place or introduced into the site including unattended items, photographers, protestors etc. Search teams may also be used to check areas prior to important meetings and to search for other items such as lost items and people. In all cases having staff trained and practiced will increase the value of the search.

Unless the site has a particular political value or is hosting VIPs protected by the State, it is highly unlikely that the police will search the premises, particularly not in response to a threat. It should also be remembered that the police are unlikely to know the site or operating environment.

Basic housekeeping such as keeping areas tidy and keeping the public to public areas makes searching easier and more effective.

Types of Search

Searches may be described as:

- **Occupant Search,** where all staff members are taught how to look around their own work areas and report anything which is out of the ordinary. Nominated individuals or teams search external areas and specialist areas such as engineering spaces. This type of search is suited to office and manufacturing sites where there is little public access.

- **Supervisor Search,** where trained wardens, supervisors or security staff search an area and ask staff or members of the public questions like "Do you know who owns that briefcase?" This type of search is suited to public areas such as shopping malls, museums, etc. and may be done surreptitiously.

- **Team Search,** where trained teams search nominated areas. This type of search is used to gain accurate information for the Threat Evaluation Team[46]. Team Search provides a high degree of confidence but requires a higher level of training.

- In addition there is **High Risk Search,** when there is a reasonable expectation that there is a bomb on the site and it needs to be found before it explodes. This level of search requires considerable training and equipment and is undertaken by trained government personnel. High Risk Search is not addressed here.

The term 'White Level Inspection' is sometimes used to describe a form of occupant search and can be presented as a requirement to look for unattended items prior to starting

---

46   For details of the Threat Evaluation Team see Chapter 4, Threats.

work each day. Given the low likelihood of a bomb incident in most work environments the concept of asking staff to begin the day by looking for a bomb may be counter-productive, creating an exaggerated sense of concern that will be replaced by a lack of effort as nothing is found. There may be value in instructing staff to use a White Level Inspection to look over their work area at the beginning of a shift to ensure a safe work place looking for and reporting items such as:

- spilt liquids,
- trip hazards,
- frayed wires,
- broken furniture,
- torn carpets, etc., and
- any unattended items.

Search Principles and Techniques

There are basic search principles and techniques, which can be adapted to suit any organization or operating environment.

The most fundamental principle is that searchers are looking for that which does not fit, that which is out of place or unusual. If the item is well disguised and integrated into the environment it is unlikely that it will be found. But, given that the item has been made in another place and then brought to the target site, and the likelihood that the person placing it will want to conceal it in some way the probability is that it will be identifiable as being out of the norm.

In addition to items that do not fit the environment, searchers should also look for any signs of unauthorised access to areas, signs of forced entry and other indicators that activities which 'do not fit the environment' have occurred.

Having staff who are trained in basic search skills and concepts will greatly enhance the value of a search. Asking people who have had no training to look for something out of place will probably waste time and resources and is unlikely to produce information of value.

Organizations should ensure that search training is appropriate to their operating environment, delivered in an effective but non-threatening manner and utilises the skills and resources available within the organization. Search training does not have to be difficult or arduous. Search can be used to find a wide range of lost or misplaced items, including children at some sites. Combined with the broader security and emergency management training, search can be presented as a useful and valuable skill[47].

---

47   See Chapter 10, Training

Personal Assistants to the Executives can be trained to look around their areas, most of which have stricter access controls, for anything out of place. They will also know if any contractors or others have been in the C-Suite area. Engineering and maintenance staff can be trained to look in the plant rooms and related spaces working from the more accessible to the more strictly controlled.

Search Priorities

The time calculation in Chapter 4 Bomb Threats will provide an indication of how much time can be spent on searching before a decision on evacuating has to be made. Given the time limit, the areas can be put into priority order. The priority will depend on the reason for the search and the number of trained searchers available and the type of search to be conducted. An occupant search is quite quick, a supervisor search of a public area can be done reasonably quickly without disturbing the public, and a team search will take longer as it is more detailed.

Priority would usually be given to:
- Areas mentioned in the threat (if any),
- Areas accessible by the public,
- Areas accessible by contractors and deliveries and visitors,
- Areas where there have been reports of or indication of unauthorised access.

Areas that are strictly controlled and where access control records can be checked should be lower on the priority list.

Concurrent activity can occur, in that personal assistants and engineering staff can be asked to check their areas while teams are searching the public spaces.

Search Team Structure

Having teams of two people works well as they can quickly divide an area, search the agreed sections and double check each other by confirming that each area has been searched.

Some large areas, such as open plan office spaces, logistic areas or production floors may need more than two people to search but care should be taken that they do not get in each other's way.

If more than one team is used there is a requirement for a Search Supervisor whose responsibility is to:
- manage the location of the search teams,
- ensure that all areas are searched,
- to check off each room/area as it is completed,

- to communicate with the Emergency Manager who is controlling the incident,
- to ensure any items that are found are properly recorded and the path to the item is marked, and
- to ensure the teams are withdrawn from the area before the deadline.

It is important that the Search Supervisor remain in a location where they can observe the teams as they enter and leave rooms or areas and from where he/she can manage their activities. The Supervisor must be careful not to get involved in searching; their responsibility is to supervise the search.

The site's Emergency Manager or other person responsible for managing the incident is responsible for:

- calculating the time available for a search,
- allocating search priorities,
- requesting reviews of CCTV, access control records etc.,
- allocating resources including teams and supervisors,
- ensuring records are maintained of all actions and decisions, and
- monitoring the time to ensure teams are withdrawn before the latest time a decision must be made.

When an unattended item is reported, the Emergency Manager will need to decide whether to continue the search or to begin an immediate evacuation. Like other bomb–related issues this will be a management decision made at the time with the limited information. If the item obviously looks like a bomb or is in the area mentioned in the threat, then the decision to evacuate is reasonably straightforward. If it is an item that is out of place and not recognised by the search team it may be reasonable to continue the search while the item is evaluated using the VALID methodology[48]. It is possible that a number of unrecognised items may be found and reported to the Emergency Manager and hence the responding emergency services. The police will determine in which order they will be dealt with depending on the descriptions provided. Basic housekeeping, tidiness and disposal of rubbish will reduce the number of unidentified items found.

48   See Chapter 5, Unattended Items

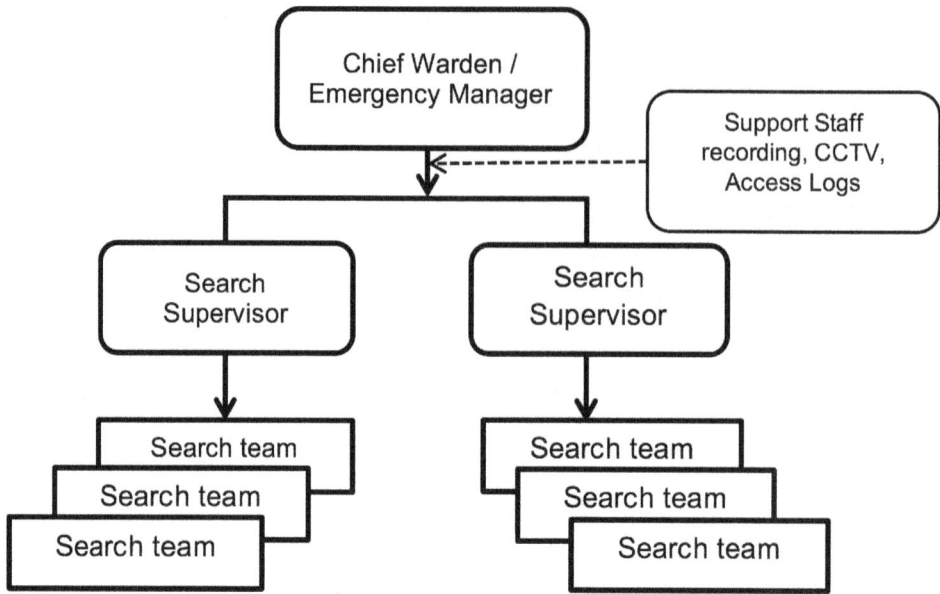

Figure 7 – Example Search Structure

Communication During Search

Ideally the use of transmitting devices should be avoided during a search. There are two main reasons:

- If the device has an electric detonator, and the connecting (or 'leg') wires are not twisted together, and they are of a length that can act as an antenna for the frequencies being generated, a transmission may cause the detonator to fire.

- If the device has a remote control firing system that uses radio or mobile/cell phone as a trigger then transmission nearby could cause it to function.

Both cases are unlikely but preventing radio transmissions can reduce the possibility. The greater risk is from hand-held radios which usually transmit on 2 or 5 watts which is considerably greater than phones. To prevent confusion by having different procedures it is suggested that all radios be turned off and phones set to "flight mode". Using flight mode means the phone can still be used to take photos and possibly to access emergency procedure apps, phone lists, etc. If the phone is not in flight mode it will continue to transmit as it seeks to make contact with the network, even if a call is not being made or received.

If radios and phones are not to be used alternate means of communication are required. These include:

- desk phones,
- building emergency (Warden Information Phones – WIP) phones,
- the use of runners to carry messages.

In some cases such as searches of external areas, radios may be the only effective form of communication. If so they should be used in open areas that have already been searched and once an unidentified item is found a minimum distance to the item should be maintained before transmitting. A range of 'safe' distances have been suggested in the references, a distance of 15 m (~50 ft.) should be appropriate for most systems. It should be noted that the distance is from the device and until the device is located transmitting should not occur. Also, the distance is spherical and includes the floors above and below although any concrete and reinforcing steel may attenuate the signal.

Search Techniques

Search is a skill that should be taught face-to-face so staff can become confident in their ability to search an area quickly, efficiently and safely. It is important that search be conducted in a calm, systematic and efficient manner. The following are basic principles that will assist in developing a search capability.

Searchers are not necessarily looking for a bomb; they are looking for that which is out of place, that which does not fit the environment.

Staff should know how to enter and move around a room so that it is searched effectively and efficiently. They must know how to identify what belongs in that environment, how to quickly search the entire area and how to report their findings. They should also know how to respond when they find an item by recording the details and by marking the route to the item for the benefit of the responding emergency services (the intent being to assist the bomb squad to deal with the item as quickly as possible).

The Search Supervisor should be allocated a clearly defined area to search by the Emergency Manager or equivalent. The Supervisor can then divide the area into smaller sections. These sections should be clearly identifiable such as a specific room, or a part of the open plan office between this clearly identified mark and that obvious location. Ideally search teams should not work next to each other. Keeping some space, like an empty room between them, is less distracting. It is the role of the Supervisor to ensure that all sections are searched, a floor plan or map will assist.

The room/area should be divided into parts and each searcher takes a defined area. The room can be divided:

- horizontally: floor, skirting board to waist height, waist to head height, above head and ceiling;
- vertically - floor to ceiling, covering the drapes, filing cabinets, wall sections, and
- in an open plan area, by volume of fittings, one person does the middle of the room, floors and ceiling while the other person does the walls and everything attached to them.

Not all divisions will be of equal size or complexity; some will have more items to search than others. Search supervisors should allow for such variations when dividing the areas and tasking teams. One search team may cover two or three divisions while another may still be working through a complex section of the room.

Common sense should be applied to prevent divisions being applied across the middle of objects. In Figure 8 the dividing line is above the books but is then lowered to the windowsill.

Figure 8 – Example of room divided horizontally

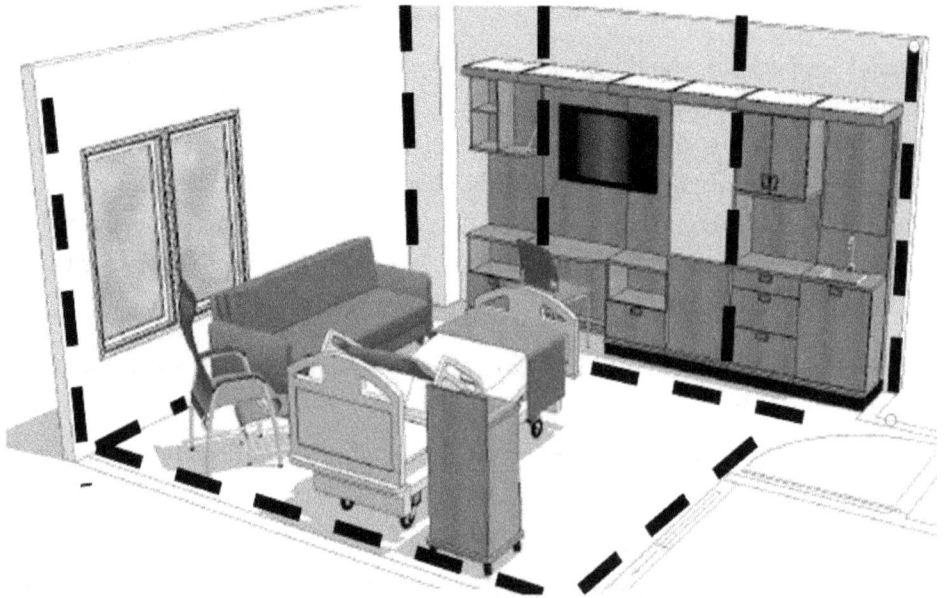

Figure 9 – Example of room divided vertically

Figure 10 – Example of dividing and open plan area

The critical part is to have a plan that ensures all items in the room are searched as quickly and efficiently as possible.

The Emergency Manager or equivalent needs to:

* determine the time available for the search and advise the Search Supervisor(s) of the time at which a decision will be made to evacuate or not,
* identify which areas are to be searched and the priorities,
* allocate resources to search the areas,
* ensure the Search Supervisors know how to contact the Emergency Manager during the search,
* manage the overall incident moving, Search Supervisors and their teams to the next priority areas as they become available.

The Search Supervisor needs to:

* know when to finish the search and whether to evacuate the teams,
* understand which area is to be searched by teams,
* divide the area into sections to be searched by the teams,
* allocate teams to the sections and monitor their progress,
* when a team reports that a section has been searched, task the team to the next section,
* maintain contact with the Emergency Manager or equivalent.

The Search Teams should:

* before entering an area, inspect the door for signs of forced entry, tampering or

anything unusual. If there is any doubt the area should not be entered until the search supervisor is notified,

- on entering the designated room/section:
  - stop at the door, listen and look around,
  - get a feel for the room and what is normal, i.e. if there is a ticking is it related to the clock on the wall,
  - is there anything obviously out of place,
  - any signs of disturbance,
- divide the room into sections and agree who will search each section,
- search every item by looking at, into, above and under,
- if cupboards, cabinets and drawers are locked and keys are not available look for signs of forced entry,
- look for signs of panels, ceiling tiles or other access points being disturbed, new hand prints or fallen dust may be indicators,
- once the room has been searched and nothing has been found, leave the room marking the door/entrance area to show the room/area has been searched,
- report to the supervisor "nothing found",
- move onto the next room/area as directed,
- if an item is identified see Actions on Find below.

The statement "nothing found" does not mean that there is no hazard just that despite a good search nothing unidentified or potentially hazardous was found. If the Supervisor and the Emergency Manager have confidence in the ability of the search teams then they will have a degree of certainty that there is no hazard. If people who are asked to search an area are poorly trained and practiced then there is little benefit in conducting a search.

If the top of a cabinet or other item can not be seen it is possible to use angles to get a better view. Team members can ask their partner "Can you see what is on top of the cupboard from the other side of the room?"

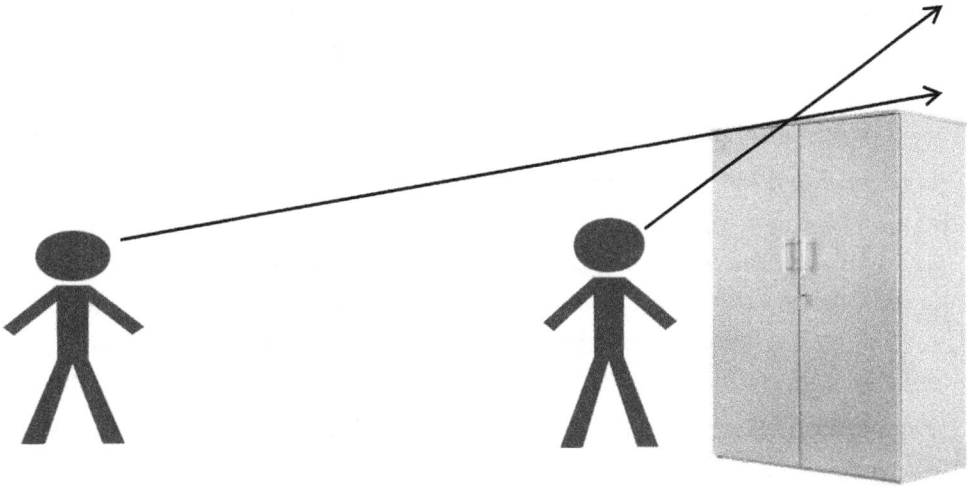

Figure 11 – Example of using angles to see above items

Obviously common sense needs to be applied. If looking for an item the size of a briefcase, there is no value in searching small desk drawers.

In some cases such as plant rooms and access spaces checking for signs of entry may be enough to indicate that no one has gained access.

CCTV does not provide the same degree of confidence as a search team but may be used to identify if there are any unusual items in areas such as empty corridors these areas should be checked by searchers but possibly as a lower priority.

CCTV can also be used to inspect external areas, evacuation routes and if possible assembly areas. Anything that is believed to be out of the ordinary should be reported to the Emergency Manager so it can be investigated by a search team.

The use of CCTV to inspect areas such as evacuation routes can begin early in the process while search teams are being organised.

Actions on Find

If an item is found during the search, decisions must be made on whether to continue the search, to further investigate the item or to evacuate the searchers and others. This decision will depend on where the item is found, why it is considered suspicious and if it matches the item described in the threat, if any[49].

If an item is found that the searchers believe does not fit the environment or is suspicious the person finding the item should:

- Not touch it.
- Inform the Search Supervisor who will inform the Emergency Manager.

49   See Chapter 9, Emergency Management for additional considerations.

- Gain as much information about the item as quickly as possible, description, location, size, shape, sounds, smells, obvious markings, holes, antenna etc. Consideration should be given to taking quick photographs with digital cameras for use by the responding emergency services.
- If directed to do so by the Supervisor, mark the location of the item, such as by using a roll of tape to mark the route from the item back towards the building/site entrance. This will assist the responding emergency services. Once the emergency services have located the end of the tape they can quickly find their way to the item, noting that the responding emergency services are unlikely to have been on the site and any technique that will aid them in disposing of the item before it explodes is of benefit.
- Report to the Search Supervisor and remain available to be a witness to the responding emergency services.

The Emergency Manager or equivalent will need to decide:
- Whether to continue the search.
- If the search is to continue, whether to continue searching in the same area as the reported item or to evacuate that part of the site. It should be noted that it is possible, if not probable, that multiple items will be reported as not fitting their environment and therefore be considered suspicious.
- Which evacuation routes and assembly area(s) should be used, taking note of the location of secondary hazards[50] and the pre-planned alternate routes.

Search Kits

If organised and practiced search teams are used on site, a basic search kit is of value. For each team a bag can be provided containing at least:
- Written reminder of how to conduct a search.
- Plans of the area(s) to be searched.
- Location of Warden Intercom Phones or other forms of communication.
- Pen and paper.
- Torch.
- Mirror with extendable handle.
- Some method of marking doors and areas as having been searched, e.g. chalk and stickers (Note, consideration should be given to not damaging the surfaces.)
- Tape to mark the route from an identified item back towards the exit.

These search kits should be stored where they can be easily accessed but protected from theft. The contents should be checked regularly.

---

50   See Chapter 9, Emergency Management for identification and consideration of secondary hazards.

Search Training and Training Aids

Search training will vary from one work environment to another and should be done in a way that inspires confidence, promotes safety and avoids alarm.

Search training aids should not look like a bomb/IED. The intent is to teach people to look for that which is out of place. Using a fake bomb can lead to the impression that this is what must be found. In many jurisdictions items that are made to look like bombs are illegal. Should someone not involved in the training find an item that looks like a bomb, considerable concern, fear and disruption will occur. Suitable training aids can be made from blocks of foam rubber, clearly marked "training aid", individually numbered and with the trainer's cell/mobile phone number. Such items are obviously not part of the normal environment, do not appear to pose a hazard and the owner can be contacted if needed. Made in different sizes and shapes they can be used in a wide variety of operating environments.

Training aids should be numbered, their locations recorded and their recovery after the training ensured and documented[51].

Summary

Search is a valuable tool in helping to assess the validity of a threat that a bomb has been placed on the site. In addition, unless a possible bomb is located there is no basis for the bomb squad to attend.

There are three types of search which an organization can employ:

- **Occupant Search,** where all staff members are taught how to look around their own work areas and report anything which is out of the ordinary.
- **Supervisor Search,** where trained fire wardens, supervisors or security staff search an area.
- **Team Search,** where trained teams search nominated areas.

Staff trained as members of a search team should be volunteers.

Search training can be delivered in a non-threatening and informative manner that shows that search is a valuable and useful skill that can be used for a range of situations.

A search capability requires a structure consisting of an Emergency Manager, Search Supervisor and a number of two-person search teams for each area.

CCTV can be used to inspect areas, particularly evacuation routes early in the process.

Basic housekeeping, tidiness and security will make search easier.

Search training aids do not need to look like bombs.

---

51   See Chapter 10, Training and Testing and selection of Consultants and Annex C for real world examples

Checklists

The following generic checklists are provided to help organizations develop their own site-specific procedures and immediate response guides.

Planning Considerations

Search - Planning	Response
Search processes and structures documented.	
Basis for search understood and documented.	
Search volunteers identified.	
Search teams and search supervisors trained and practiced.	
Communication systems between search teams, supervisors and threat evaluators documented, available and practiced.	
Training aids appropriate and well managed.	
Search kits available and appropriate to the site.	

Incident Checklist

Search - Action	Response
Reason for search understood.	
Time limits on search defined, documented and provided to search supervisors.	
Communication and reporting protocols reinforced to searchers and search supervisors.	
Areas to be searched identified in priority order.	
Search teams and search supervisors allocated to areas.	
Search teams report progress via appropriate communication systems.	
If item(s) found apply Unattended Item Processes to determine if they pose a hazard.	

Search - Action	Response
Search teams withdrawn before deadline.	
Actions documented.	

Chapter 9

Emergency Management Considerations

Emergency Plans

A bomb incident raises additional emergency management considerations to those normally provided in fire-oriented emergency procedures. Most organizations have some form of Emergency Control Organization (ECO) with one person responsible for managing any emergency that may arise. This person may be the Emergency Manager, Incident Controller, Chief Warden, Security Manager or have some other title. In this Chapter the term 'Emergency Manager' is used.

If there is a bomb on site, or the threat is assessed as being credible, or the unattended item is believed to be hazardous then the correct response is to move people away from the hazard, i.e. to evacuate. The evacuation arrangements should be specified in the site's Emergency Management Plan.

A key consideration is that moving occupants, if there is no hazard, exposes people to risks of crush, falls, extreme exertion (in the case of the unfit), external weather and other hazards. Evacuation should only be initiated if there is reason to believe that a bomb or other hazard may be on site. In the case of a fire, the identification of the hazard by a witness or by an automated alarm system makes the decision simple. In the case of a threat or unattended item the decision to evacuate becomes one for management, after due consideration of the facts[52].

Of particular concern are the templated or 'Insert client name here' Emergency Management Plans that do not accurately reflect the functions, occupants or nature of the facility. Many of these are provided by suppliers with little regard to the actual building with only the evacuation diagrams differing from one site to another. Generic plans do not recognise that there may be specific populations such as a childcare or health centre on site nor do they tend to identify primary and secondary hazards. Often they do not consider the implications of having underground car parking or whether the site is sole or multi-tenanted.

Such plans and related procedures, when reviewed objectively, are found to be incomplete, potentially dangerous and certainly leaving the Emergency Manager, tenant and site owner exposed to claims of failing to provide a site-specific emergency response capability. Emergency Managers, Facility Managers and Security Managers should customise any generic Emergency Management Plans to ensure they accurately represent the hazards, occupants and response options to be found at the site.

---

52   See Chapters 4, Threats and 5, Unattended Items

If the organization is a tenant on a site with inadequate emergency plans there may be no ability to implement their own detailed emergency plans. What the organization can do is develop additions to the site plan that provide supplementary considerations and response measures for their own areas and staff.

The Emergency Management Plans should link to and be compatible with other management plans such as Security, HR, BCP, Environmental, Media and Legal. A lack of coordination between the plans will result in unnecessary delays and disruption.

Evacuation Considerations

In response to a bomb, distance and cover behind a solid structure provide the greatest safety. Evacuation assembly areas should be:

- At least 100 meters (preferably 300 m if achievable) from the building to provide some protection from blast, fragmentation and heat effects, see additional comments on evacuation distances, below.
- Out of direct line of site of the bomb.
- Not facing or under windows which could be subjected to blast effects.
- Behind solid cover such as another building.

The evacuation should begin with those closest to the item.

Technical information relating to the 100m recommended minimum distance is provided at the end of this Chapter.

Depending on the size of the device and its location consideration may be given to a partial evacuation. Staff may be safer remaining in the workplace if they are some distance from the item and behind substantial cover. If the device is small and able to contain no more than about 1kg (2.2 lb) of an explosive material an evacuation of the floor with the item and one above and below may provide adequate safety.

A factor relating to partial evacuations is people's peace of mind, if they see others being evacuated they may wish to evacuate as well. The relevant manager may decide that although those further away would be safe it may be better for reputational and HR issues to evacuate the site. The decision may relate to the level of training and awareness of the staff.

All Emergency Management Plans should have allowances for alternate evacuation routes and assemble areas which should preferably be on opposite sides of the site so that the one chosen on the day is up wind of smoke and other hazards. The ability to select from alternate evacuation routes and assembly areas will also make it difficult for a bomber to determine the exact response to an incident. During fire drills fire wardens[53] and staff

---

53   Depending on the jurisdiction, 'Wardens' may be referred to as 'Marshals' or other titles.

can be practiced in using alternate evacuation routes.

The security of assembly areas must be considered, as threats may be used to move people out of secure areas to less secure assembly areas. There may be a need to balance the use of less preferable but more secure assembly areas during periods of heightened risk. Evacuation routes and assembly areas should be inspected by fire wardens or others to ensure they are safe to use, clear of all obstructions and with no unidentified items. This adds a degree of protection against an 'ambush device' in the assembly area. CCTV may be used to inspect external areas, evacuation routes and if possible assembly areas. Anything that is believed to be out of the ordinary should be reported to the Emergency Manager so it can be investigated[54].

If the bomb is in another part of a large site, consideration may be given to keeping people inside, based on 'Shelter-in-Place' procedures[55]. This will depend on the relationship between the buildings, the size and location of the bomb, the structural strength of the building and similar factors. The decision will be a balance between the exposure presented by evacuating people and the risks of leaving them in place. If people are kept in buildings, they must be moved away from any windows or glass facing the direction of the bomb and should use intervening walls as added protection.

Effective Emergency Management plans must not only consider how people will be removed from a hazard but also how they will be protected once evacuated. In particular, the aged, ill, infirm and very young need to be protected from the elements, the effects of the hazard and possibly from other people.

How and when neighbouring businesses will be informed that an evacuation is being initiated should be pre-planned. There is a moral, and probably legal, obligation to inform those in close proximity to a potential hazard. Consideration should be given as to whether the organization or the responding emergency services should inform others. It should be noted that an effective and practiced threat evaluation capability may help protect against any claims from others that they were needlessly inconvenienced by the decision to evacuate or were put at risk because of a decision not to evacuate.

Emergency response considerations should address the following:

- Assembly areas that permit the required number of staff to be held in comfort, out of the weather and preferably with access to amenities.
- Does the site have groups of aged, mobility or intellectually impaired or very young as regular occupants or as visitors?
- The ability to evacuate access control and visitor records to the assembly area so a check of all those that are recorded as being on site can be conducted. This can

---

54   See Chapter 8, Search

55   For example see Lee 2007

possibly be done through the use of remote IT access.

- A communication system, message centre number, web site or other method of informing staff of the situation and providing updates.

- Depending upon the time of day, consideration may be given to letting staff disperse and return later or the next day.

- If people are to be dispersed can they safely get to their vehicles or will other means of transport need to be provided and if so, by whom. In some more remote sites arrangements may need to be made with bus companies to move the people to the nearest public transport.

- If there is a public car park on-site what are the implications for people not having access to their vehicles - particularly if the evacuation was a management decision rather than one required by the Emergency Services?

- The ability to 'evacuate' information and other critical assets but not at the risk to life.

- A 'Shelter-in-Place' procedure, i.e. the ability to keep people inside if the threat is external, and should consider access to washrooms, food, water and communications.

Emergency Management Plans should include procedures for safely closing down processes to reduce loss to the organization and to minimise hazards. Some manufacturing processes may take considerable time to close down safely; while leaving them operating may create potential hazards to the site. In some cases doing a quick close down to assist an emergency evacuation may cause considerable damage to the equipment, possibly resulting in permanent loss of capability. On sites where food is prepared, consideration should be given to closing down cooking fires, gas lines etc. to reduce the hazards. Where possible, closure or monitoring of processes should be conducted from remote sites.

In some industries, such as petro-chemical processing plants, it may not be possible to evacuate the control room without generating greater hazards than those posed by the bomb. In such cases, the control rooms are often designed as bunkers offering a high degree of protection from external events.

The fallacy of opening doors and windows

Some emergency procedures include the requirement to open doors and windows if the evacuation is related to a bomb. This is unnecessary for the following reasons:

- Opening doors and windows will take time that would be better spent moving people away from the item.

- In modern commercial buildings the windows cannot be opened, and those that can often require special tools.

- The device would have to be quite small for the blast to vent through doors and windows without doing considerable structural damage.

- If the device is large enough that it will remove the window frames etc. then opening the windows and doors will be of very limited, if any, benefit.

Emergency Rendezvous Point

Consideration should be given to the location of the emergency rendezvous points (ERP) where the responding services can be met and briefed. The people at the ERP should include the witnesses who took the call or found any item, a senior manager, the threat evaluation team leader, the Emergency Manager and preferably the site engineer or facility manager. Accurate site plans in either hard or soft copy, including the location of secondary hazards, should be brought to or be available at the ERP to assist in briefing the emergency services and to help them plan their response. Prompt, accurate information will help the emergency services save the organization's facilities and possibly lives.

Secondary Hazards

Secondary hazards are those materials on-site that are safe until acted upon by an explosion. Managers need to know the type and location of all hazardous material and processes so emergency services can be briefed; such information may be held in HAZMAT plans. Some hazards, such as high pressure oxygen or water lines, may not be classified as hazardous material but can still be secondary hazards if they are damaged by an explosion. Knowing what secondary hazards could be affected by an explosion will assist in deciding the correct evacuation route and assembly area. Secondary hazards can be marked on a map of the site that should be accessible to responding emergency services. Not all secondary hazards are stationary, some like mobile fuel trailers can be relocated and some may be temporarily on site.

Some processes involve hazardous materials that must be carefully managed to prevent accidents. The site's emergency plans should address such concerns in relation to immediate evacuations and the consideration that staff may not be able to return for some time.

Briefing the responding Emergency Services

The responding Emergency Services, particularly the bomb squad, if attending can be expected to ask the following:
- Where is the item?
- Who found it (or received the threat)?
- Why the item is considered hazardous?
- What it looks like: size, shape, smell, visible wires, switches packaging, markings, smoke or burn marks?
- Are there any photographs of the item?
- If the item is under CCTV can the image be accessed without entering the building

containing the item?

- Access to the item via the most direct and alternate routes?
- What other hazards are near the item?
- What processes are currently operating (i.e. for a manufacturing site) and what hazards do they pose?
- What has been done to the item (if anything)?
- What has been done to evacuate people?
- Who and where is the Emergency Manager or equivalent?
- Who is in charge of the site and capable of making executive decisions?

The organization should be ready to provide this information.

Hazardous Mail Response

For Hazardous Mail incidents the guidance in Chapter 7 Hazardous Mail should be applied and reflected in the Emergency Management Plan.

Security during and after an evacuation

If the building is on fire there is little likelihood of someone trying to gain access but if an evacuation has been initiated because of a threat or unattended item the site may be vulnerable. It is important to ensure that the site is secured during and after the evacuation.

If the site is insecure and some of the entrances are not visible from the evacuation assembly area it is possible that people may enter the site without realising it has been evacuated. The Emergency Management Plan should address: how will doors be secured to ensure no one can enter the site and what signage is available to indicate the site is closed?

Most emergency egress doors fail 'safe' meaning that they remain unlocked. It may be possible to have the doors lock as long as those seeking to leave can open them from the inside.

Managers may need to consider how vehicle egress will be managed and how will traffic be stopped from entering the site?

Reoccupation

A resolved bomb incident, if there is no explosion, may result in immediate reoccupation of the site. The following considerations should be included in the Emergency Management Plan:

- How the site will be reoccupied and who will make the decision? The responding Emergency Services may provide advice but the final decision is likely to rest with the organization's management.

- Who will be allowed in, when and how?

- How will the assets be protected during reoccupation, particularly if there is a large public population on site i.e. a shopping centre?

- Should security staff be amongst the first to enter to provide guidance and surveillance as others move back in?

- Will a search of the site, or selected areas, be conducted before reoccupation? This may be of benefit if a credible threat was received but no bomb was found before the evacuation.

Post-Blast Considerations

After an explosion the emphasis will be on immediate first aid, evacuation and securing the parts of the site not affected by the explosion, with later implementation of the organization's BCP. See Chapter 6 for additional considerations relating to post-blast scenarios.

Recovery of and aid to those caught in the rubble of a collapsed building may be beyond the scope of the organization.

The site's Emergency Management Plan should include, or point to other plans that provide, guidance on deploying first aid, maintaining control over the area, initiating the BCP, advising families, and responding to legal and media requirements.

These response capabilities are required for all incidents where multiple casualties and/or structural damage may occur and therefore should not be written purely as "post-blast" rather they should be generic "post-incident" plans.

After Hours

All of the above considerations should have an after-hours alternative. When the site is not fully occupied, many of the resources will not be available and the only people on site may be security guards or a few shift workers.

Media Management

The organization's Media Management plan should include prepared scripts to present the required messages to staff, stakeholders and the public in response to a range of scenarios including a bomb incident. Important themes may include:

- that the organization has suffered an incident (threat, unattended item, bomb possibly found and disrupted by the emergency services), or is in a post blast situation,

- that the organization has a plan which was implemented to protect, as far as possible, staff and others,

- that the organization is assisting the emergency services with the response and any investigations.

It is possible that an explosion may have been the result of an industrial accident or other cause. Unless it is obvious that a bomb was involved, the cause of the incident should be left for the investigating agencies to confirm.

Guessing at the motives behind the incident should be avoided.

Social Media

Staff should be advised, in prior training, that recording and sending images of any sort of accident or incident is inappropriate and could result in their being prosecuted, depending on the jurisdiction, or of having their camera confiscated by law enforcement as evidence. If possible, pre-arranged screens can be used to block site lines to the scene.

The Media Plan may wish to address if staff are free to comment or post on social media that they are safe, to reassure friends and relatives who may otherwise come to the site looking for them, or clog inquiry lines asking after them.

Technical details on Evacuation Distances

Table 2 shows the approximate blast distances for various charge sizes. At the stated distances the blast will drop below 34.5 kPi (5 psi) which is given as the pressure at which there is a chance of a minor injury such as an ear drum rupture[56]. Due to a number of factors such as the position of the person in relation to the blast, reflective surfaces and variations in the explosive charge these figures can only be indicative.

Charge Weight (TNT equivalent)	Eardrum threshold distance (34.5 kPa)
5 kg	<10 m (33 ft.)
10 kg	<13 m (43 ft.)
23 kg	<17 m (55 ft.)
100 kg	<27 m (89 ft.)
225 kg	<35 m (115 ft.)
500 kg	<46 m (151 ft.)

Table 2 – Example blast distances

56 UFC 3-340-02 Figure 1-3

The figures in Table 2 support the requirement for a **minimum evacuation distance of 100 m (~330 ft.)** and the ability to shelter behind something solid. A solid barrier will also assist in protecting against fragmentation, which can be expected to travel further than the blast wave. See Chapter 1 for additional information on fragmentation effects.

Problems with common evacuation distance guidance

Government agencies provide evacuation distance guidance[57]. These evacuation distances are provided as indicators of safe distances and represent open-field environments where a person in the open at the specified distance would be well beyond any blast effects and fragmentation trajectories. They are not applicable in urban environments and can not be applied effectively. As an example, the ability to quickly and efficiently evacuate people to a distance of one and a half kilometres or more is beyond most organizations' capabilities.

Table 3 provides an extract of common evacuation guidance from various agencies containing similar net explosive quantities (NEQ)[58]. Included in the table is an indication of the size of the area to be evacuated. These distances are not applicable to most organizations.

Type of Bomb	NEQ	Preferred evacuation distance for the public	Resultant Evacuation Area
Pipe bomb	500 g (1.1 lb)	860 m (~2800 ft.)	~2.3 square km (~0.9 square mile)
Briefcase/Suitcase	23 kg (50 lb)	1,520 m (~1 mile)	~7.26 sq. km (~2.8 sq. mile)
Small car	230 kg (~500 lb)	1,915 m (~1.2 miles)	~11.5 sq. km (~4.4 sq. mile)
Small truck	4,500 kg (9920 lb)	3,280 m (~2 miles)	~33.8 sq. km (~13 sq. mile)

Table 3 – Extract from Evacuation Guides and showing resultant evacuation area

The historic or intelligence-led basis for nominating these NEQ as standard or common explosive charge weights has not been identified. These NEQ appear to be 'worst case' scenarios, where the figures of what is 'possible' rather than 'probable' are provided. This

---

57 For example US National Ground Intelligence Centre, Improvised Explosive Device Safe Standoff Distances Cheat Sheet

58 NEQ is the weight of the explosive excluding packaging and fragmentation casing, see NATO AAP-6 (2007)

has significant implications for the manager planning an evacuation.

These evacuation distances are impractical in all but open field and low population areas. The population density of London[59] is given as 5100 per km^2, for New York as 2050 per km^2; the working population would be higher. The recommended evacuation distance for the 500 gram pipe bomb would probably evacuate a substantial part of the business district of a city.

Managers in urban environments should consider the practicality of the information provided and their ability to implement it. They should plan how to evacuate their people away from the hazard, for a bomb to a distance of **at least 100 m (~330 ft.)**, out of sight of the location of the bomb and behind solid cover. If a greater evacuation distance is required the emergency services will coordinate it.

Summary

Bomb-specific additions to Emergency Management Plans will also be of benefit in relation to other emergency situations.

Evacuation for a bomb should be to a distance of at least 100 m (~330 ft.), out of sight of the location of the bomb and behind solid cover. If a greater evacuation distance is required the emergency services will coordinate it.

Evacuation plans should provide adequate distance and shelter for a bomb incident as well as for a fire.

Ensure the plans include an emergency rendezvous point where the appropriate people can meet with and brief the responding emergency services personnel.

Consider at what point the neighbouring businesses should be informed, particularly in a multi-tenanted building or site. It can be quite disconcerting for nearby tenants to see others evacuating but not know the reason why. Delay in evacuating other tenants may also place them at unnecessary risk.

Consistency between the various management plans will provide a safer and more effective capability to respond to an emergency.

---

59   www.citymayors.com/statistics/largest-cities-density-125.html

Checklist for Briefing Emergency Services

The location of the item?	
Who found it (or received the threat)?	
Why the item is considered hazardous?	
What it looks like: size, shape, smell, visible wires, switches packaging, markings, smoke or burn marks?	
Are there any photographs/images of the item?	
If the item is under CCTV can the image be accessed without entering the building containing the item?	
Access to the item via the most direct and alternate routes?	
What other hazards are near the item?	
What processes are currently operating (i.e. for a manufacturing site) and what hazards do they pose?	
What has been done to the item (if anything)?	
What has been done to evacuate people?	
Who and where is the Emergency Manager or equivalent?	
Who is in charge of the site and capable of making executive decisions?	

Chapter 10

Training and Testing and Consultants

Training Considerations

Bomb safety and security should be the responsibility of everyone on site. When selecting a training provider, either in-house or external, ensure that the training offered is sensible, applicable and reflects the operating environment of the site.

Training can be via the organization's intranet, posters and handouts, 'tool-box' talks, or via formal face-to-face sessions. A variety of methods may be required to reach all staff.

The training should be

- Relevant to the person's work environment.
- Realistic, neither alarmist nor trivial.
- Delivered in a rational, non-threatening manner.
- Based on the organization's published policies and procedures.
- Be simple to understand and implement.

It is important that training and testing improve the organization's capabilities and develop a positive attitude towards bomb security.

The effectiveness of the policies, procedures and training should be tested via exercises from desktop exercises for managers through to practical exercises for those responsible for search, assessments, and response measures.

Exercises should be aimed at reinforcing the staff's confidence and skills.

All Occupants

All staff, tenants and contractors should know how to: **R**eceive, **R**ecord and **R**eport unattended items, threats and what looks like a bomb. All people on site also need to know how to identify and respond to hazardous mail.

Visitors and the public should be able to report items of concern to staff with confidence that the report will be treated seriously and that the staff know how to refer the concern to the relevant manager.

Staff should be advised, in prior training, that filming and sending images of any sort of accident or incident is inappropriate and could result in their being prosecuted, depending on the jurisdiction, or of having their camera confiscated by law enforcement as evidence.

Management Training

Managers should be trained in how bomb incidents are managed, and the potential impact on the site, the organization and the corporate reputation if incidents are not well managed. Selected managers will need to be trained on how to assess (**Review**) threats and reports of unattended items including how to gain more information though witnesses, CCTV, access control records, search or other means

Managers also need to know what **R**esponse options are available.

Managers should understand the implications of evacuating or deciding not to evacuate and develop the skills to determine when an evacuation is necessary.

Desktop exercises can be used to introduce and practice managers in the required capabilities leading to management-level oversight and decision making during practical search and evacuation exercises.

Bomb incident training and testing can be used as a lead in to practicing Evacuation, BCP, Media Management, and other plans.

Threat Evaluation Training

Managers on the Threat Evaluation Team (TET)[60] need to be trained in the processes and procedures required to assess threats. Threat assessment training can range from small desktop exercises to larger scale exercises that involve the deployment and management of search teams and bomb-specific evacuation drills.

When conducting a threat assessment exercise it is recommended that the existing situation at the site be used to reflect the operating environment. For example the time, weather, VIP visitors, number of people on site, maintenance activities should all reflect what is happening at the time so as to make the threat assessment as realistic and relevant as possible.

The assessment of a threat will be related to the knowledge within the TET. It is possible that one group, having specific knowledge of the security measures or other factors, may decide that it is not possible for the person to have done what they claim. Another TET provided with the same information but with a lower level of knowledge may decide that it is feasible. Neither is incorrect as long as the decision is based on the best information available and can be justified. Threat scenarios can be drafted to minimise the possibility of conflicting results.

Decisions that may be of concern are if during the evaluation the trend is towards the fact that it is not possible for the person to (for example) have put a bomb in the CEO's office but the decision is made to evacuate anyway. Or if, due to maintenance work, security

---

60   See Chapter 4, Threats

has been relaxed and it is feasible that people could have gained access but the decision is made not to search or respond because of the fear of causing offence or disruption. If the assessment does not match the recommended response there is a disconnect that needs to be addressed.

Training of threat assessors should also include the recording of the assessment, the decisions made and the basis behind the decision.

Security and Emergency Staff Training

Security and emergency staff have specific responsibilities and skill sets. They are the ones most likely to be informed of a threat or unattended item and must know how to respond.

They may need to be trained in assessing unattended items.

They should know how to select and implement an appropriate response if they believe a hazard exists. They should also understand the implications of initiating an unnecessary evacuation.

They may also need training in search techniques and to understand why searches are conducted, what is expected and how to search an area so they provide a high degree of confidence that the area was well searched.

Security and emergency staff need to understand how bomb incidents fit into the broader emergency management plans and how a bomb incident may differ from a standard evacuation.

Security staff should understand how the site is to be secured during an evacuation, while the site is empty and during reoccupation.

Security and emergency staff should also know how to establish an Emergency Rendezvous Point and who needs to be there[61].

Both desktop and practical exercises will help them develop the required skills.

Training Aids

While some awareness of the types of systems that can be used to trigger an bomb may be of interest and value to staff, care should be taken to ensure that the training does not become a 'how to build a bomb' or "These are bombs I faced in my career" lectures. All training, including on the types of bombs that may be encountered, must be relevant to the site, the occupants and the function of the facility.

As stated in Chapter 8, search training aids should **not** look like a bomb. The intent is to teach people to look for 'that which is out of place'. Using a fake bomb can lead to the

---

61   See Chapter 9, Emergency Management

impression that this is what must be found. In many jurisdictions items that are made to look like bombs are illegal. Should someone not involved in the training find an item that looks like a bomb, considerable concern, fear and disruption will occur[62]. Suitable training aids can be made from blocks of foam rubber, clearly marked "training aid", individually numbered and with the trainer's cell/mobile phone number. Such items are obviously not part of the normal environment, do not appear to pose a hazard and the owner can be contacted if needed. Made in different sizes and shapes they can be used in a wide variety of operating environments.

Training aids must be controlled. Each should be numbered and their location clearly documented, possibly including a photograph. Upon completion of each training scenario the items must be recovered and documented. At the end of the training all training aids must be counted and any missing items recovered.

Consideration must be given to what will happen if a training aid is lost or discovered by a person not involved in the training. The use of innocuous items will minimise the likelihood of unnecessary disruption.

When training staff how to report and describe an article the accuracy of the description of a block of foam rubber or other innocuous item is of value. Part of the training should be for the person to describe the item found, the item can then be compared against the description. Again, it is stressed that the item does not need to look like a bomb for this to be effective training on observing and describing the item.

First Aid Training

There are specific considerations for first aid responders in relation to blast injuries:
- Not all blast wounds immediately apparent.
- Record details for all present with particular note of those who felt the blast wave.
- Blast and fragmentation injuries are evidence and any fragments should be retained along with details of how they were recovered.
- Bodies are evidence and, after being checked to see if first aid is required, should be left where they lie.
- Depending on the number of casualties, there may be a need to initiate a triage capability.

If the risk of a bombing is considered likely, consideration should be given to seeking specialist professional training from those with experience with blast injuries.

Selection of Consultants

When selecting a consultant to assist with bomb safety and security planning, assessments

---

62   See examples in Annex B and Annex C for real world examples

or training, managers should apply similar standards as they would for any other managerial-level advisor

- Appropriate, formal qualifications in the discipline in which they are consulting.
- Relevant experience related to their area of expertise.
- Membership of relevant professional associations.
- Recognised independently issued certifications.
- Relevant insurances.
- Where applicable, relevant licences or registrations.

Consultants in the area of bomb safety and security tend to fall into a number of categories

- Those with a background in emergency or security management but who have little or no understanding of blast effects.
- Those with a military or law enforcement background with experience in dealing with bombs or the intelligence related to them but with limited understanding of business imperatives.
- Those with no understanding of the emergency and security management disciplines or bomb-specific knowledge.
- Those with appropriate qualifications and experience and who understand the business drivers.

Operational experience dealing with bombs in conflict zones and urban environments is of benefit and provides a significant technical knowledge. It is the ability to apply that experience to the corporate sector that is of relevance to the client.

The most important consideration is that the consultant be able to apply their knowledge and skills in a manner that is relevant to the organization's built and operating environment. A consultant working to the public or private business sector should understand the necessity to protect reputation, functionality and profitability as well as life.

When defining the scope for a bomb incident consultancy tender or contract, care should be taken to ensure the requirements of the organization are specified and will be met, rather than being limited to and reflecting the consultant's background and experience.

Interoperability of Plans

Bomb safety and security training should not be conducted in isolation. The opportunity should be taken to ensure the bomb-specific elements are aligned with the site's Emergency and Security Plans. Bomb training also provides the opportunity to test the BCP, Media, HR and related plans to ensure they are compatible and coordinated.

Summary

Training in how to Receive, Record, Report, Review and Respond appropriate for staff positions within the organization and their level of responsibility is essential if a safe and effective bomb incident management capability is to be established.

Training and testing can be delivered in non-threatening, informative and relevant ways that will enhance people's skills and understanding.

Chapter 11

Risk Assessment and Mitigation

Security and Emergency managers are often required to explain their concerns and to justify their requirements based on a risk assessment. The risks relating to bomb incidents can be assessed and mitigation strategies developed using standard security risk management principles and techniques[63].

This chapter reiterates and expands upon information and guidance provided in previous chapters.

This chapter provides guidance on defining the risks in relation to the various types of bomb incidents. An assessment of the likelihood of an organization or individuals being exposed to the risk is based on a threat and vulnerability assessment that encompasses the changing profile of the organization and the protective and procedural security measures in place.

The consequences of a bomb incident can be assessed and rated based on knowledge of the assets at peril and the potential effects of the bomb incident under consideration, given the operating environment of the organization.

While recognising that the initiative rests with the perpetrator, as does the design and placement of a bomb, it is possible to identify and implement appropriate risk mitigation treatments. These treatments will usually increase the ability to deter a bomb incident (Likelihood) and to detect that a bomb incident has or will occur and to have appropriate response measures in place to respond to the incident and to minimise the harm to the assets and operation of the organization (Consequence).

This chapter also provides guidance on defining the risks, the factors that influence the likelihood and consequences, the security risk management analysis factors to be considered, and risk mitigation options for each of the types of bomb incidents (bombs, unattended items, threats, mail bombs and post-blast).

Context – Built and Operating Environments

Risk assessments must always consider the natural, built and operating contexts in which the risk may be realised. In relation to bomb incidents this can be addressed in terms of the built and operating environments of the site.

The built environment includes: the physical construction, the surrounds, the distance to the perimeter and degree of control over that space, landscaping, neighbouring structures, utilities, as well as on-site and nearby hazards. The operating environment encompasses

---

63 For example ISO 31000, Australian Standard Handbook (ASHB) 167 Security Risk Management and the Security Risk Management Body of Knowledge.

what the site does rather than what it is. The operation of the site addresses what occurs including the critical areas, functions and controls as well as security factors such as policies, procedures and resources used to prevent, detect and respond to explosive attacks.

The operating environment also encompasses the 'image' of the site. If the site is an open, welcoming, family friendly venue it will have considerably different operating characteristics and business drivers than a high-security corporate or government facility where access of people and items can be tightly controlled.

Most organizations are unlikely to be subjected to a bombing, but as suggested in Chapter 1, the motive, material and knowledge exists. The relevant manager, trained and experienced in security risk assessment, should assess and monitor the likelihood of the risks relating to bomb incidents.

Risks Relating to Bombs

The principles outlined in this chapter apply to all types of bombs: hand delivered, vehicle-borne, mail bombs, suicide bombs etc. The detailed application of the principles will vary depending on the threat analysis, exposure and vulnerabilities of the organization. Separate sections are provided in this chapter on the specific considerations for mail bombs and suicide bombs.

While it is not possible to predict what quantity or type of explosive will be used in a bomb, it is possible to make assumptions based on the layout and functions of the site. There is a direct relationship between the nature/function of the site, the motive of the attacker, the intent of the explosive attack and hence the design of the bomb[64]. A realistic threat assessment with consideration of the tenants, neighbours, functions, operating and built environments will assist in determining the probable targets and explosive attack vectors.

In some cases a vehicle-borne IED (VBIED) may be a likely attack method. In other cases, due to access controls, stand-off distances, screening and the nature of the site a VBIED may be the least likely attack method with placed, projected, suicide or mail bombs having a higher probability.

Consideration of the built and operating environments will assist in identifying what can be placed where and therefore what are reasonable risk assessments and mitigation measures.

The risk that a perpetrator may have the motive and means to build and deploy a bomb is not one the organization can readily manage. Rather, the risks related to a bomb being used against the organization can be defined in terms of:

* *Failure to prevent a bomb from entering the work place ...*

---

64   See Williams D. 2003.

- *Failure to protect assets from a bomb (on site/off site/when travelling)...*

These can be further devolved to:

- *Failure to identify a bomb ...*
- *Failure to respond appropriately to a bomb ...*

Factors that will influence the likelihood assessments for such risks will include:

- The existing social, political, workplace and other issues which might make the organization a more likely target for an act of violence. These could be determined from a specific bomb vulnerability assessment or as part of an enterprise-wide assessment. It is essential that bomb incident risks be reviewed as the operating environment, exposure and vulnerabilities change. It is possible that the organization's efforts and relationships elsewhere may increase or decrease the likelihood of an attack and that the likelihood of a bomb incident will change over time.
- Basic access control measures, including adequate boundary protection, access systems ranging from electronic systems to key control processes, and restricting the public-to-public areas.
- Security in depth, providing additional levels of control and restricted access to higher value assets.
- Good workplace practices such as keeping areas tidy and clean so that any items introduced to the work area will be quickly identified.
- Staff awareness and an ability and willingness to identify and report items which are out of place.
- Supervisors/managers having the training and knowledge to respond appropriately to a report of a bomb from staff or the public.
- The proximity of assets to the boundary, e.g. when the building housing the asset forms part of the boundary and the bomb can be placed near the external wall. In such cases the likelihood of detecting such an item and the ability to respond to it should be assessed. CCTV and mobile guards may be factors in this assessment.
- Detection equipment. If bomb detection equipment has been deployed it is important to ensure that it is appropriate and capable of the task for which it is employed, that the equipment is deployed as part of a cohesive security plan, that it is maintained, that staff are trained and that processes are in place to respond if they find that which they are looking for, i.e. a bomb[65].

It will be noted that most of these factors are those required to maintain a normal safe and secure work area. Therefore in many respects mitigating the risk of a bomb incident will build on existing security and safety measures.

---

65 See Chapter 12, Physical Protective Considerations

The consequences of failing to prevent a bomb include:

- An explosion should the bomb detonate, in which case the specific consequences will vary depending on the construction, type and location of the bomb, particularly its proximity to various assets. Response considerations for an explosion are outlined in Chapter 6 Post-blast.

- Disruption to operations should the bomb be identified before it detonates, including while an evacuation is initiated and the incident is managed by the site's Emergency Control Organization. The best protection from a bomb for staff and visitors is usually distance provided by an evacuation. The site's evacuation plan should be reviewed to ensure that the evacuation distances are far enough away from the site to provide adequate protection. Many fire safety evacuation assembly areas are too close for bomb safety considerations.

- Concern by staff, clients and public over the handling of the incident. Documented and rehearsed procedures will assist with this issue.

- Hazards faced by people during an evacuation such as road crossings, egress for children, aged or those with disabilities should be addressed in the site's emergency procedures.

- Depending on the motive and skill of the perpetrator, there is a possibility of other bombs being on or near the site. The emergency procedures should include the requirement for the egress routes and assembly areas to be inspected for hazards prior to or during the evacuation. Such an inspection is good practice as it will also detect blocked exit routes or other barriers and hazards[66].

See also consequences relating to Post-blast, below

Even though the likelihood may be assessed at the lower end of the scale, the consequences of an explosion can be catastrophic and appropriate risk mitigation treatments should be considered.

Suicide Bombers

The management of risks related to suicide bombers requires the same considerations i.e. identifying the hazard and having responses in place. Organizations need to carefully consider if they are likely to be the target of a suicide bomber, which represents a very small sub-set of bombing perpetrators. The identification and response to suicide bombers may be more difficult as there may be less time to respond as the suicide bomber *"is the ultimate guided weapon capable of changing its target and timing".*[67]

The technique for identifying a suicide bomber is the same as for other types of bombs,

---

66 See Chapter 9, Emergency Management

67 C Williams. Suicide Bombers. Australian Homeland Security Research Centre presentation 2005. Additional information on suicide bombers' motives and techniques can be found in Williams C. 2004 , Bloom M. 2005, Pape R. 2005.

recognising that which is out of place, and that which does not fit the environment. Detecting a suicide bomber may be difficult as they usually disguise or conceal their bombs and can alter their approach if they think they may be detected. If a suspected suicide bomber is identified, usually the best way to minimise the consequences is to immediately start to move people away and to limit the bomber's ability to get closer to the organization's assets. Experience shows that once a suicide bomber believes they have been detected or is to be thwarted in their attack, they will try to detonate the bomb to cause as much effect as possible.

Vehicle Bombs

VBIEDs pose particular problems in relation to assessing and managing risks. In respect to likelihood a VBIED can be positioned as close to the site as common traffic routes permit, sometimes as close as the width of a sidewalk. If the site has public under-building parking then a VBIED can be introduced into the site without any screening procedures.

In relation to consequences a VBIED can be assumed to be carrying a larger quantity of explosives than a PBIED. The larger charge weight can cause greater damage from a greater distance, possibly outside the area under control of the organization. The effect on the site will be considerable, particularly if it occurs under the building.

Loading docks also create an exposure, which can be exploited to introduce a VBIED into the site. Controls can be introduced to ensure that delivery vehicles are pre-authorised reducing the likelihood of a VBIED entering the loading dock. The consequences can be managed by identifying which critical functions and utilities are located in or near the loading dock and if they can be moved or hardened to a reasonable and realistic degree[68]. For high-security sites, ideally all delivery vehicles should be required to attend an off-site loading dock and the goods cross-loaded to in-house transport but this is an expensive option.

Risks Relating to Unattended Items

There is a reasonable probability that unattended items will be found at any organization. It is possible the item may be lost or left goods, workers' toolboxes, tourists' bags, rubbish, courier-deliveries or a bomb. Such items are more likely in public areas such as foyers, entranceways and outdoor areas. In most cases there will be doubt as to the nature and origin of the item. If an item found on site obviously appears to be hazardous i.e. visible wires and components that look like explosives, switches etc. or because of its location i.e. strapped to fuel tank, then the item should be considered to be a bomb and the appropriate measures implemented.

The aim is to determine if the item poses a hazard. The risks related to unattended items

---

68   See Chapter 12, Physical Protective Considerations

could be defined in terms of:

- *Failure to identify an unattended item …*
- *Failure to respond appropriately to an unattended item …*

The probability of detecting unattended items increases if the staff are trained and aware and willing to identify and report such items to supervisors/managers. Additionally supervisors/managers should know how to action such a report including the ability to assess the item using a methodology such as VALID[69]. The likelihood of finding rubbish or abandoned items on site is higher than that of finding something that appears to be a bomb. The origin of the item may be confirmed through a review of CCTV, access control records, interviewing staff and others, asking for the owner to return etc. (see examples at Annex C).

Some sites assessed as being at high risk, e.g. international airports, have portable X-ray systems to assist with investigating unattended items. For most organizations this will not be a cost effective capability.

The consequences of failing to respond appropriately to an unidentified item can be considerable. If the process is mismanaged the primary consequences are:

- Failing to evacuate when there is a real hazard.
- Unnecessary evacuation and disruption when a simple investigation would have shown that there was no hazard.

The consequences of finding an unattended item can be mitigated if processes are in place to identify how the item got there, what it contains and if it poses a hazard.

The response procedures for unattended items should include what will be done with the item if it is not hazardous, e.g. passed to the lost property office, disposed of, returned to owner etc.

If there is any suggestion that the item poses a hazard, then the risk alters to that of having a bomb on site and the pre-planned procedures should be implemented.

Risks Relating to Bomb Threats

A perpetrator needs to spend little time or other resources in making a bomb threat against an organization. The organization has little ability to prevent a threat being made by phone, e-mail, SMS, mail, fax and other communication systems. The ability to recognise, evaluate and respond to a threat lies within the capability of all organizations regardless of size or function[70].

The instinctive reaction whenever a threat is received may be to evacuate although

69   See Chapter 5, Unattended Items

70   See Chapter 4, Bomb Threats

this may not be the safest or most sensible action and does not reflect security risk management principles. Constant evacuation will undermine the employees', clients' and owners' confidence in management's ability to provide a safe, secure and productive work environment. Constant evacuation will also lead to 'copy-cat' incidents as staff seek time off work or outsiders enjoy the prospect of disrupting activities. The perpetrator will also learn that the organization always reacts in the same way and can capitalise on that routine behaviour. At the other extreme some organizations have an (unofficial) policy of not responding to threats as 'they are always hoaxes'; this approach is both negligent and dangerous. Bomb threats can be subjected to risk management practices.

The risks relating to bomb threats (and also for other types of threats against the organization) could be defined in terms of:

- *Failure to respond appropriately to a bomb threat …*

The appropriate response is based on the ability to capture the information about the threat and analyse it. The question is not "Is the threat real?" as the threat has been received and is a real threat. The question can be reworded as "Is it feasible for the perpetrator to have done that which they claim?" This is a question that managers should be able to answer based on their knowledge of the site's security and procedures.

For some organizations the number of bomb threats peak at certain times such as in educational institutions where they tend to be more common at exam time than during holidays. Some industry sectors such as the aviation industry receive many threats in a year and deal with them on an almost routine basis. Other organizations are rarely if ever threatened, it is these that are at greater risk of failing to evaluate and respond in the safest and most appropriate manner.

The likelihood of a risk, such as that suggested above, being realised will depend on the organization's ability to:

- Recognise they have been threatened.
- Capture and report the information about the threat to the relevant authority within the organization.
- Evaluate the threat.
- Respond appropriately.

If it is determined that the threat is not credible then work may continue but procedures should be in place to record the evaluation and the basis for the decision and to inform all those that are aware of the threat, of the decision.

If it is believed that the bomb threat is credible, in that the perpetrator may have done that which they claim, then the emergency procedures for responding to a bomb should be implemented. If time permits, a search of the area can be conducted on the basis that if the item is found the bomb squad can be deployed with the aim of rendering the item safe prior to its exploding and damaging the assets. Again there is a need to record the evaluation and the basis for the decision.

The consequences of failing to respond appropriately will vary from failing to evacuate when there is a real hazard to unnecessary evacuation and disruption when evaluation of the threat may have shown that it was not necessary.

How the organization handles bomb and other threats will be reflected in the confidence of staff, clients, owners and the public in the ability of management to deal with such risks.

Risks Relating to Hazardous Mail

The mail and courier systems offer a degree of anonymity to the perpetrator and the ability to deliver the bomb or other hazard directly to the targeted individual[71].

The initiative for constructing and sending hazardous mail rests with the perpetrator and the organization can do little to modify this. Therefore the risks relating to hazardous mail may be defined in a similar manner to other bomb incidents:

- *Failure to identify hazardous mail ...*
- *Failure to respond appropriately to hazardous mail ...*
- *Failure to prevent hazardous mail from entering the workplace ...* (If the mail is received and sorted within the work place this is a difficult risk to manage.)

The likelihood of failing to identify hazardous mail is directly related to the training and awareness of all staff who receive and open mail. Guidance on detection needs to be applied to the particular operating environment of the organization.

The consequences of failing to identify hazardous mail will include injury and possibly death of the person who opens the mail. Other consequences include disruption while investigations are conducted and the area is repaired.

The ability to respond appropriately to hazardous mail is dependent upon considered, appropriate procedures relevant to the operating environment of the organization. The procedures should cover:

- Reporting of the suspected hazardous mail.
- Investigation as to why is it considered suspicious.
- Determining if it is considered hazardous or safe to open.
- If it is considered hazardous, appropriate responses based on the type of hazard.
- Evacuation considerations for some or all of the site.

The investigation into the items can include:

71   See Chapter 7, Hazardous Mail

- Comparing the item to mail bomb recognition posters and similar guidance.
- Asking the recipient if they are expecting the item.
- Checking information on the sender, if any.

Similar risk considerations can be applied to courier-delivered items with the difference that courier-delivered items could contain a timed triggering mechanism and therefore if a couriered item is suspected of being a bomb, an immediate evacuation of the area should be initiated.

Some organizations transfer the risk by having a contractor receive, inspect and open the mail. In such cases it is of value to ensure the service provider has appropriate response measures in place to minimise disruption to the mail delivery system.

Risks Relating to Post-blast

If a bomb is not detected, despite the best plans and measures, and there is an explosion the risk of a 'post blast' situation should be considered. A post-blast plan is usually considered part of the consequence management treatments, but it can be defined as a separate risk in terms of:

- *Failure to respond appropriately to a post-blast situation ...*

As an explosion is one reason why an organization may lose access to its site, information or people, the post-blast plan must be aligned with the Business Continuity and Business Resumption Plans[72]. A post-blast consequence management plan will address the following:

- The main consideration will be a staff support function in that a bombing is a deliberate human act of violence and the psychological and societal effects on the staff and others may be considerable.
- If there are casualties the organization's HR support plans will be required.
- The scene of the explosion will be a complex crime scene and may be isolated by the investigating authorities for a considerable period of time.
- Urgent structural assessments of the building may be required.
- The ability to obtain repair services, parts and equipment maybe restricted, particularly if the explosion affected a number of buildings in the area.
- Insurance may not cover the bomb damage depending on the exclusion clauses in the policy, in which case other means of funding repairs and replacements may be needed.
- Legal protection from litigation or claims of negligence may be required. A demonstrated adherence to bomb incident security risk management practices may assist with a legal defence.

---

72   See Chapter 6, Post-Blast

It may be that the bomb was not targeted at the organization and that the organization suffered 'collateral damage'. If the incident is external the organization should have a 'Shelter in Place' plan so that people can remain where they are until the nature of the hazard is identified and the most appropriate egress route and time for an evacuation can be determined. If the building is damaged then people should be evacuated in the safest manner.

Similar consideration to those above can also be used for planning consequence management for accidental explosions due to industrial accident etc.

Summary

The various types of bomb incidents can be assessed and mitigated where appropriate using security risk management principles. Knowledge of the motives behind the use of bombs and bomb threats, the effects of explosives, the existing security environment, including the training and awareness of staff, are required to accurately assess and treat bomb incident risks.

Chapter 12

Physical Protective Considerations

Physical Measures

In addition to the managerial policies, procedures and practices there are a range of physical measures, which can be implemented to reduce the likelihood and consequences of bomb incidents. These are:

- Distance,
- Detection,
- Response,
- Strengthening/Hardening.

Some of these measures have been mentioned in previous chapters and are expanded upon here. These observations are generic and each site will need to be assessed and specific treatments applied where they are cost-effective and consistent with the image and functions of the site and tenants.

Many of the physical measures may already be in place, as they address a number of security risks, hazards and threat vectors. If they are in place, it may be necessary to ensure that they are applicable to bomb safety and security by verifying that the associated procedures, polices and training also address the identification and response to bomb incidents.

Bombings are in most cases a low likelihood/high consequence risk so bomb-specific measures should be subjected to a cost benefit analysis[73]. The value of implementing bomb-only measures may not be economic but if they are aligned to the detection of and response to other hazards, then they may be justifiable. For example search technology may be used for the detection of other weapons, drugs, alcohol, protest material, other prohibited items, or to detect corporate property being stolen.

If seeking advice on bomb protective measures, clients should ensure the consultant or provider has a demonstrable understanding of blast effects as they apply in the urban environment and that the proposed solutions actually address the identified risks[74].

In all cases physical measures must be supported by appropriate policies, procedures and training. For equipment such as X-ray machines and metal detectors there are defined competencies in their use. It is a managerial responsibility to ensure the in-house or contracted staff operating the equipment have the relevant training. Their use must also be in accordance with local regulations.

---

73   See Stewart (2008)

74   See also Chapters, 10 Training and Testing and Consultants and 13, Modeling

If the intent of the physical protective measures is clearly understood then the effectiveness of the process for selecting and installing equipment and other measures can be monitored and assessed.

The Function of the Site

The key planning factor is the nature and function of the site. If it is a site where the public are free to occupy such as shopping centres, public parks and general thoroughfares, then it is difficult to apply restrictive barriers and processes. In such cases the emphasis is likely to be on non-intrusive observation, detection and response rather than seeking to prevent a bomb from entering the site. In countries where the bombing threat is immediate and constant, strict controls over who and what can enter public access areas may be justified.

For sites where the public is encouraged to attend and the image of the venue is an open, welcoming and friendly environment but where some control over who enters is possible, additional screening and control measures can be implemented. This is particularly true for events that are 'ticketed' or where access is granted under specific conditions. In these cases it may be possible to restrict what items are brought into the site and to implement searches of people and goods.

At sites where access is granted only by permission, higher levels of control into and through the site can be maintained. For sites where a high degree of access control is maintained not only can a screening capability be implemented, it may be expected, for example at international airports where transit between groundside and airside zones is controlled and monitored. In such locations it is expected that goods and people will be screened, particularly the public, who have not been through a vetting/approval process.

Areas within the structure vulnerable to bombs should be subjected to specific protective consideration such as hardening of walls and strengthening of supporting columns. Such areas include foyers and other public access areas, loading docks, mailrooms (and any work area where mail is opened), external areas and carparks be they open air, high-rise or underground.

In all cases the protective measures must support the image of the site that the occupant wishes to portray.

Distance

The most practical form of protection from a bomb is to prevent it from entering the site and to keep it as far from the site as possible[75].

A key factor is the geographic boundary. In most cases, physical protective measures can only be applied to the limit of the site's boundary. This boundary may be the outer

---

75   See Chapter 1, Introduction and Explosive Effects

perimeter fence, the door to the building, the entrance to the lift well on the floor occupied, or even just the office door.

The aim should be to allow in only those people and items that are known and permitted to enter the site or specified area. Prescriptive controls and barriers that do not support the function and image of the site are counter-productive and will not be acceptable to the organization's owners.

External Standoff

An effective way to keep bombs away from buildings is to have a large site with a perimeter fence and controlled access points where all those entering, goods and vehicles can be searched. This option is often not available or suitable for most organizations. Where it is applicable, assurance is essential that adequate screening mechanisms, processes, training and capabilities are in place otherwise the process is flawed and of limited benefit.

For most sites with control over their external perimeter, keeping a bomb away from the outside of a building is usually only applicable for vehicle borne bombs. Vehicle access can be denied or restricted through the use of bollards (rated to a suitable impact level) or through landscaping. There are a number of guides on designing and implementing barriers using aesthetically pleasing landscape, plantings, street furniture and art works[76].

Rising or other barriers across driveways are of benefit if the approach is under observation and the barrier can be activated before a suspicious vehicle enters the site.

When designing vehicle inspection points it is important to allow for a turnaround area for vehicles that are denied entry. This turn around area must be positioned so the vehicle does not have to enter the site to return to the road.

In many cases barriers will prevent cars and trucks but not bicycles, motorbikes or scooters from accessing the site. In such cases the amount of explosives that can be brought to the building has been reduced from (potentially) 100s to 10s of kilograms with resultant reduction in damage and consequences.

When considering external perimeters it should be noted that a bomb does not actually need to enter the site to damage facilities near the fence line. Defence in depth is a basic principle of security whereby a series of access controls are placed between the outer entrances and assets of value. Defence in depth is only partially applicable in relation to bomb security, as a bomb does not need to be in a room to damage the contents. The blast wave and associated effects resulting from a bomb external to the building (including debris from shattered walls and windows) can enter and cause considerable damage.

If critical equipment, functions or utilities are near an external fence then consideration

---

76   For example: New York City Police Engineering Security Protective Design for High Risk Buildings, FEMA 426 and CPNI Protecting Against Terrorism

should be given to who can park or access the other side of the fence. If controls over access and parking on the other side of the fence are not possible, then the ability to either move the asset or harden the intervening wall should be assessed[77]. Effective surveillance and an ability to respond to investigate any activity on the boundary may assist.

Internal Standoff

While efforts should be taken to prohibit IEDs and other weapons from entering the site, it may not be possible to prevent the public and others from bringing unscreened items into a site. Therefore, measures should be taken to keep items that have not been assessed away from critical assets and functions.

One method that can be effective is the use of cloakrooms. Aligned with published policies about what may be brought beyond the entrance area, the cloaking (temporary storage) of goods can act as a deterrent. The perpetrator, once aware of the requirement, may be less willing to place a bomb in a cloakroom rather than the targeted area. Although handbags and other smaller items will probably still be permitted, the size and hence the resultant damage of a bomb that can be brought further into the site has been reduced. The size of a bag inside the controlled area can be reduced from a wheeled bag (< 23 kg) to a one-hand carry (~ 5 kg).

If there are public car parks on the site, the proximity of critical assets and utilities should be identified. If possible, public parking near these areas should be prohibited and the parking areas either removed or allocated to trusted employees and some degree of access control, possibly including removable bollards, be used to prevent unauthorised parking.

These principles are based on effective access control keeping unauthorised people away from critical and valuable assets and functions.

If there is limited control over access then consideration can be given to strengthening or protecting the critical utilities, such as not running the main communication lines and utilities past areas where an unscreened person could place an explosive device.

VBIED Security Considerations

Security considerations to limit the effect of vehicle bombs, depending on the role of the site, include:

- Limiting public parking on site.
- Limiting parking close to the site to employees and other trusted personnel.
- Removing or limiting parking near critical utilities and assets, or moving those assets and functions away from parking areas.

---

[77] A related risk to the site is an incident (deliberate or accidental) that results in a vehicle crashing through the fence or exterior wall. Again, control over the parking area or hardened fence or walls will mitigate this possibility

- Physically hardening critical utilities and assets close to public parking.
- Installing vehicle access barriers.
- Using fences, bollards, trees, terrain features or other barriers to limit vehicle access to the site other than through controlled entry points.

In cases where higher levels of security are required, the following may be considered:

- Pre-registration of drivers and vehicles.
- Barriers at the inspection points suitable for stopping forced entry, possibly with a second barrier behind the vehicle to prevent tailgating.
- Detailed inspection and detection procedures, including appropriate measures if the driver, vehicle or load are of concern.
- An isolation bay to the side of the inspection point, but outside the barrier to enable the vehicle and driver to be segregated while additional inspections are conducted. This may need to be hardened or separated from the sites critical utilities or assets.
- A turn around area, outside the barrier, for vehicles that will not be permitted access.

For many sites, particularly those in city centers, these measures will not be applicable in which case consideration should be given to hardening the perimeter of the site.

Detection

Ideally a bomb should be detected before it can detonate allowing time to evacuate people, initiate the BCP and call the Emergency Services. Detection technology is constantly evolving but the following considerations may assist in selecting and deploying a detection capability.

Clearly defining what is prohibited and why is an essential step in developing a screening policy and selecting appropriate technology. What is prohibited will depend on the nature and function of the site - prohibited items might include weapons, dangerous goods, protest material, alcohol or recording equipment. In other cases these items may be essential to the operation of the site. What is prohibited should be clearly displayed as a 'Condition of Entry'. If items are prohibited the site should have the ability to detect such items and have supporting procedures for what to do if they are detected, i.e. confiscation and disposal, temporary holding and return to owners upon exit or refusal of entry. Consideration should also be given to what is to occur if items are detected that are classified as illegal under local legislation.

Depending on the 'image' of the venue or activity, the screening capability can range from having all people and goods pass through screening equipment to having trained staff observe those entering and asking to inspect any items or people that appear out of place or otherwise not normal to the environment. These capabilities must be in accordance with published Conditions of Entry and with local legislation.

Trained and aware staff provide one of the best detection capabilities. Staff who know how to identify that which is out of place or does not fit the environment, and are empowered and encouraged to report such items and people, with a supporting response capability will provide an effective ability to detect potentially hazardous items and activities.

CCTV is a useful technology, if it is constantly monitored and is linked to an immediate response capability. Developments in monitoring systems allow for detection of items that remain stationary or people moving in specified locations. Again, processes for constant monitoring and immediate response are required to support such capabilities. In the absence of an immediate response, the primary function of CCTV is recording what happened for later evaluation.

The principles of Crime Prevention Through Environmental Design (CPTED)[78] consider aspects such as lighting, view lines and colours to make it easier to identify items and people that are out of place. CPTED should be a factor during site design.

Specialist Inspection Equipment

Specialist inspection equipment can be used, where appropriate, to search people, vehicles and/or goods entering defined areas. In all cases the ability and limitations of the equipment should be understood.

Planning for screening capabilities includes considering floor space, adequate space for queuing, electricity, security staff areas, holding lockers for prohibited items and possible separation areas for people whose entry might be delayed. A key consideration is 'what will happen when we find what we are looking for?' For each detection capability there needs to be associated response measures to ensure the event is recorded, reported, contained and controlled with the minimum possible disruption to the site's functions and while remaining compatible with the organization's image.

Metal detectors are designed to indicate if there is any metal content within the screened item. Metal detectors can be walk-though, hand-held, or belt fed. The presence of metal does not mean a hazard exists so additional screening is required once an initial alarm has been given.

The ability to look into an item without opening it enables the detection of most hazardous items. X-ray machines come in a range of sizes and types ranging from small portable items that can be linked to digital image capture screens to cabinet X-rays and belt-fed systems similar to those seen at airports. The volume of items to be processed will determine the size and capacity of the machine, for example a small cabinet machine may be of benefit to those organizations with few people seeking access. For those requiring a greater flow rate, a belt-fed X-ray will probably be more appropriate. Organizations may

---

78   For example see http://www.cpted.net/

find additional uses for an X-ray machine once they have one in place, for example the ability to inspect some equipment or the non-intrusive screening of briefcases prior to sensitive board meetings. There are WH&S implications relating to the use of X-rays, and organizations must ensure the X-ray machines meet all legal requirements and are maintained by qualified and certified agents. Effective use of X-ray machines requires staff to be trained and certified in X-ray interpretation and possibly in X-ray operation.

There are specialist detection systems designed to detect bombs and other items placed under vehicles. Bombs placed under vehicles are usually designed to kill the driver or passenger[79]. A device still attached to the underside of the target vehicle when it enters or leaves a site suggests that either it failed to function on start up or is on a timer or other delayed trigger. Under-vehicle bombs tend to be small and are unlikely to do significant damage to the building. The purpose of installing an under vehicle screening system needs to be defined. Is it to protect VIPs and others who may be at personal risk, is it to prevent small bombs from entering the site or is it to detect other types of prohibited items that may be secreted under a vehicle?

There is a range of explosive detectors available. Most work by taking a vapour or physical sample and subjecting it to a chemical analysis. These systems are designed as investigative tools where the presence of explosives is suspected and confirmation is required. Explosive vapour detectors (EVD), which are based on analytical technology such as gas chromatography, are relatively expensive and require staff to be properly trained in their operation. Some have the advantage that they are dual technology and can also be used to scan for illicit drugs, which may be of benefit to some workplaces. EVDs may be cost effective where an explosive threat is considered a likely event. In all cases procedures should be developed to determine if any explosives detected are the result of legitimate use such as from those that work with explosives, pyrotechnics or ammunition. A means for verifying such claims of employment must be established.

Strengthening / Hardening

Structure

A degree of protection from blast effects can be achieved by strengthening the building. Strengthening can limit the ability of blast effects to enter the building and reduce the likelihood of collapse as a result of an explosion[80].

Walls and supporting columns can be hardened through a number of techniques including encasing columns in steel jackets to deflect blast and reduce the blast's shattering impact on the concrete. Walls can be faced with protective layers to absorb blast and fragmentation impact. Additional columns and supports can be added to provide strength.

79   See Chapter 3, Bombs

80   See Yates 2012

During construction or refurbishment the use of steel fiber reinforced concrete, or similar additives, will improve the blast resistance of structural elements. The increased use of design methods to prevent progressive collapse should one or more supports be compromised also assists in limiting the effects of an explosion.

A consideration when considering blast is that the force will be exerted upwards against floor slabs. Floors may be designed for a downward loading of say 10 kPa being 4 kPa dead and 6 kPa live loads. When considering blast there may be an upward force measured in hundreds of kPa. Similarly side loadings from an explosion are both considerably larger than those for weather events and are applied in a fraction of the time; milliseconds as opposed seconds.

For every site, detailed advice from engineers and other professionals knowledgeable in blast effects will be required and the type of strengthening can be related to the expected charge load, anticipated distances and confidence in the security measures to keep larger devices away. It is important that the design team and engineers have a sound understanding of blast and its effects otherwise inappropriate, ineffective and overly expensive protective measures may be recommended and implemented[81].

Glazing

There has been considerable research into the ability of glazing to withstand blast effects[82]. Glazing can be strengthened through the use of films, replacement with specific glazing materials and through capture systems such as blast curtains and retaining wires, although these can detract from the look of the building. The development of blast resistant glazing systems is on-going and research into current best practices and products is recommended. When considering how glazing may have been hardened, the manner in which the windows are secured to the building should also be addressed. The probability of the entire window and frame being blown out should be addressed as the window may be fitted to light weight frames with a 'bite' (the amount of glass held within the frames) of only a few millimeters[83].

The international standard for testing of glazing material against blast is ISO 16933 (or ISO 16934 depending on the test method). The usual measure of damage is the USA General Services Administration "Standard Test Method for Glazing and Window Systems Subject to Dynamic Overpressure Loadings" which rates window failure on how glass enters a defined space.

The use of protective film on windows is to reduce the hazard of glass fragments caused by explosions which are a primary cause of injuries, particularly to the eyes. Blast protective

---

81   See also Chapter 13, Modeling

82   For example: Netherton 2012

83   See UK CPNI and US FEMA and UFC publications relating to glazing.

films can reduce the fragments from Toughened Tempered Plate (TTP) glass or spalling/delamination of laminated glass[84].

Laminated glass is preferred in terms of blast protection. The UK Centre for Protection of National Infrastructure (CPNI) has issued a number of papers related to blast resistant glass. CPNI states *"The minimum overall thickness of laminated glass classed as blast resistant is 7.5 mm, including a minimum polyvinylbutryal (pvb) interlayer thickness of 1.5 mm."*[85] The US Federal Emergency Management Agency (FEMA) states *"Laminated Glass is the preferred glazing material for new construction."*[86]

The New York Police Department (NYPD) in their guidance on protective design for high rise buildings states[87] : *"Treated window glazing can incrementally increase the blast resistance capability of glass. Although no commercially available glazing can fully mitigate the effects of a close-range blast event, certain glazing systems may substantially reduce blast impact at greater distances. Window glazing can also reduce the distance that glass fragments travel upon failure."* And, *"... buildings (should) avoid the use of annealed glass completely, and limit the use of fully thermally tempered glass to windows on upper floors, where increased distance from street level reduces potential blast pressure and glass fragmentation danger. For windows on lower floors, the NYPD recommends ... buildings use laminates."*

The effectiveness of the film is related to the fixture methods for the panes. The film will stop the glass from fragmenting but has limited effect of preventing entire panes from failing and being projected into the building. Security film is usually fitted to the inside of the window surface and the method of adhesion to the frame can assist in securing the glazing to the frame. The strength of the frame fixture to the structure and the surrounding panes will dictate the ability of the panels to withstand a shock front. Ideally the framing system should exceed the strength of the glazing so that the window will fail and, as a laminate with security film, break into large shards that will be retained in the frame, or being larger fragments will strike fewer people.

Containment

In some cases it has been argued that rubbish bins should be removed altogether. While removal may address the issue of a bomb being hidden in a bin, the question needs to be asked if all other places of concealment such a shrubbery, street furniture, signage etc. should also be removed. In addition, removal of bins may increase hazards such as infection, infestation, trip hazards, damage to the environmental image of the site and the visual impact caused by discarded rubbish. If a bomb in a bin is considered likely, other options include frequent emptying of bins, locating bins away from key areas, and

---

84   Spalling is the release of fragments from the inside face of a surface due to the transfer of the shock wave through the material.

85   Reference: http://www.cpni.gov.uk/advice/physical-security/ebp/blast-resistant-glass/

86   FEMA 426 Reference manual to Mitigate Terrorist Attacks Against Buildings section 3.4.5

87   NYPD Engineering Security Protective Design for High Risk Buildings, 2009

smaller access holes to limit the size of what can be placed in a bin.

Containment bins that can either contain the blast effects (of a specified size) or direct the blast are offered for sites where the public need to dispose of rubbish. If containment bins are used the amount of explosive force that can be contained must be noted and compared to what is a reasonable risk for the site, remembering that up to 5 kg can comfortably be carried in one hand without undue stress. If bins that direct blast are used, their positioning is important as placing such a bin under a large concrete roof will cause the directed blast to be reflected back down with potentially greater effect than the initial pressure.

In relation to containment systems for HAZMAIL, a number of factors need to be considered. As per the guidance in Chapter 7, items containing what is suspected to be a chemical or biological hazard should not be moved once they have been opened and the material released. Rather the items should be covered in place to reduce the spread of the contaminant. If an item is identified as possibly containing a chemical or biological hazard before it is opened then a containment system may be of benefit. As a general principle, items that are suspected of containing explosive or incendiary material should not be placed in containers as this may increase the fragmentation effect and may make it more difficult for the responding Emergency Services to examine and render-safe the item.

Response

Physical response capabilities include the communication systems and the ability for those on site and the responding Emergency Services to limit the effects of a bomb.

Access to CCTV images of the area or item of concern from a remote site, such as the emergency rendezvous point[88] will be of great benefit. This capability allows ongoing observation and decision-making after the site has been evacuated.

The ability to communicate with all on site to inform them of the incident and the appropriate response actions is essential. Communicating with the Wardens and other Emergency Control staff, and also with the tenants, visitors and public, is normally addressed through the site's Emergency Management Plan. The ability to communicate with neighbouring buildings is often not addressed and the technology to develop a precinct communications strategy should be considered. In some venues the electronic 'way signing' and advertising boards can be used to issue emergency messages and guide evacuation routes.

Communication with search teams should also be considered[89]. The use of desk phones,

---

88   See Chapter 9, Emergency Management

89   See Chapter 8, Searching

Warden Intercom Phones, runners and other non-transmitting systems are alternatives to radios or mobile phones. In some cases, such as searches of external areas radios may be the only effective means of communication in which case they should be used sparingly and at least 15 m in every direction from any unattended item.

The storage of the site's floor plans and other essential documents so they can be accessed off-site is of benefit to the responding emergency services. Evacuation plans should address how plans can be stored off-site, where they are accessible 24/7, or how relevant computer-based files can be accessed off site.

Summary

While the primary protection from bomb incidents is managerial, physical protective measures can assist in deterring bombers, making identification easier, assisting with response measures and in lessening the effects should a bomb detonate.

Chapter 13

Considerations for Modeling Blast Effects

Introduction

The effects of an explosion are a legitimate concern for architects, designers, engineers, owners and tenants and they will seek guidance as to how bombs may affect the building, both the physical structure and the functions conducted on the site.

A useful tool is computer modeling. This chapter provides observations on types of modeling available and considerations when seeking blast modeling.

The modeling of blast informs both the design of buildings yet to be constructed and the strengthening of existing structures.

A key factor is that all assumptions related to the use of bombs as an attack vector are inaccurate as IEDs are by definition 'improvised'. It is not possible to predict what type and quantity of explosive will be used, how it will be primed or detonated, how it will be encased or when and where it will be detonated. Intelligence-led information, consideration of potential motives and attack vectors and the protective security measures in place can assist in determining probable bomb attack methods.

It is important to ensure that the effects modeled:

- are based on realistic attack vectors,
- relate to the actual built and operating environment,
- provide information that is of value and applicable.

To achieve a useful outcome the request for modeling must be based on a clear understanding of what question is to be answered and of the limitations of blast modeling.

Another consideration when modeling explosive effects is an acceptance of the fact that should a bomb detonate, there will be damage and probably death and injury. If the bomb is large enough or placed in close proximity to a critical area, the consequences to the site and the organization may be catastrophic.

Built and Operating Environment

The design of the site, aligned with appropriate policies and procedures can limit, within certain parameters, where a bomb can be placed and this can have a direct relationship to the survivability of the site[90]. Modern building design often considers the possibility of progressive collapse should one or more structural elements be removed.

Consideration of the built and operating environments will assist in identifying what type

---

90   See also the Context section of Chapter 11, Risk Assessment and Mitigation.

of bomb can be placed where and therefore what advice can be obtained via modeling. A considered assessment of what type of explosive attack and the probable NEQ relates to the types of explosive charges used as the basis for design, modeling and testing.

Areas within the structure vulnerable to bombs and close to critical areas could be subjected to blast modeling. These include foyers and other public access areas, loading docks, mailrooms (and any work area where mail is opened).

Methodology

When considering what a bomb could do to the site, a methodology to assist in defining the problem is to:

- Consider the design, function, image and assets of the site and its surrounds.
- Identify critical areas and features in the structure and the existing security measures limiting access to, and near, them.
- Determine reasonable explosive attack vectors for delivering explosive devices to the site, PBIED, VBIED, Hazardous Mail.
- Consider the probable composition of explosive devices including the type of explosive and reasonable explosive charge weights. This informs the subsequent blast effects calculations and observations.
- Use modeling to calculate overpressure and impulse effects to enable a review of the structural integrity of the building design from both internal and external explosive attack.
- Use modeling to calculate overpressure effects on critical areas within the building.
- Identify physical and procedural protective measures to mitigate vulnerabilities.

Type of IED

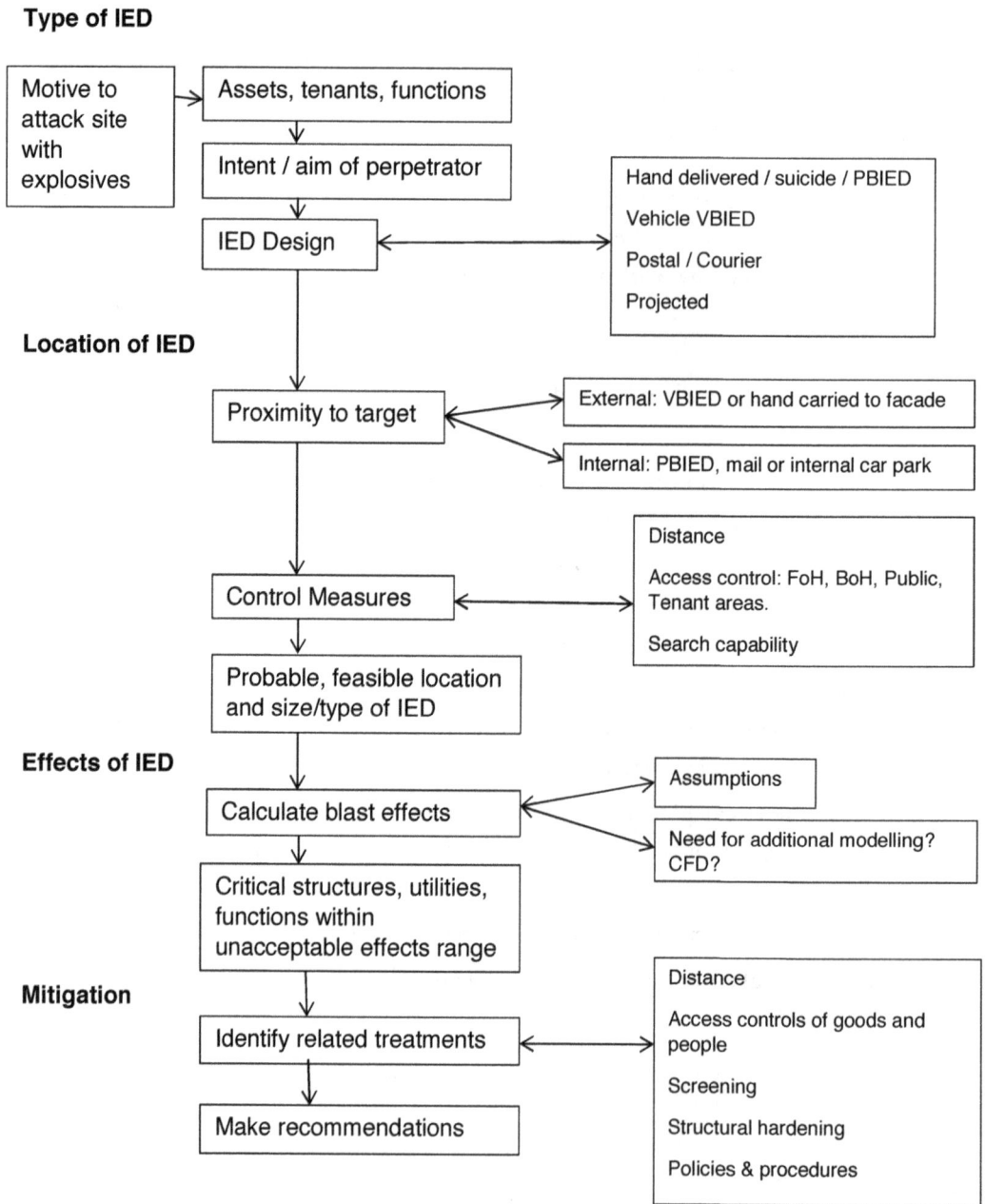

Figure 12 – Blast Assessment Flow Chart

Blast Calculations

The base equation for explosive effect is the 'scaled distance' equation $R/W^{1/3}$ where R = range in metres (m), W = explosive weight in kilograms (kg). As the explosions are assumed to be at ground level on a resilient surface, the effects are given as an expanding hemisphere and hence blast pressure dissipates on a cube root basis.

In relation to structural damage the following generic failure pressures are quoted:

- 48 kilopascals (kPa) or roughly 7 pounds per square inch (psi) which is given as the upper threshold for *"severe damage to steel framed buildings"*[91].

- 690 kPa (100 psi) at which *"Most construction materials will sustain major damage or failure at these peak pressure levels"*.[92]

These generic figures can be modified to reflect the actual construction of the site including the compressive strength of the concrete, the type of reinforcing, etc.

Building designs are based on expected static and live loads and local environmental conditions including wind strength, earthquakes, storms, forest fires, etc. Commercial buildings are usually not designed to withstand explosive events.

Explosive effects are orders of magnitude greater than normal design criteria i.e. blast can result in 100s or 1000s of kPa delivered in milliseconds as opposed to wind loadings which are usually given as single digit kPa delivered over seconds. Buildings which are specifically designed to withstand explosive over-pressures are explosive storage bunkers, military command centres and process control rooms in places such as petro-chemical plants and are not, usually, aesthetically pleasing.

Types of Blast Modeling

The modeling of blast effects can be described in numerous ways. For example, Remennikov describes three methods: Numerical (or first principles), Empirical (or Analytical) and Semi-Empirical[93]. While these are technically accurate the simple divisions provided below are considered appropriate for this Chapter.

There are, in fundamental terms, two levels of blast modeling:

- Detailed Computational Modeling.
- Conventional Modeling.

Detailed Computational Modeling. The blast effects of an explosion are analysed in terms of gas and hydro-dynamics. This modeling can utilise computational fluid dynamics (CFD) or finite element analysis (FEA) to model the impact of the pressures and impulses so the resultant responses of the structural elements can be plotted[94].

Detailed computational modeling provides detailed results but requires a high degree of input as to the construction of the structure and surrounds as well as a large amount of computing power.

In some cases such modeling may produce results which are too detailed. The exact manner in which a particular structural member may fail, given a specific charge size and

---

91   FEMA 428

92   FEMA 426

93   Remennikov 2003

94   Examples of such computer models include AUTODYN and LS-DYNA.

location, may not be as relevant as the fact that it **will fail** if a bomb above a certain charge weight is placed within a specified distance.

Conventional Modeling. Conventional modeling uses software tools such as CONWEP (Conventional Weapon Effects Calculation)[95] or proprietary systems to calculate blast pressures from base principles. CONWEP is validated against empirical studies[96] and is approved for use in assessing blast loading on structures[97].

Conventional modeling generates simpler, less-detailed results than detailed computational modeling but often results in more relevant information as it presents a broader indication of the type of effects that the structure will experience. This information provides managers with the ability to begin physical and procedural protective and response measures.

Selection. The selection of detailed computational modeling or conventional blast modeling will depend on what information is being sought. If the potential impact of an explosion against the site from a given distance is required then conventional modeling is probably appropriate. If the exact manner in which a structural element or façade will fail if subjected to specified overpressures is required then detailed computational modeling may be necessary.

Modeling Considerations

Modeling may state that specific results are expected based on the defined criteria; to gain full benefit from the results it is necessary to recognise the limitations of the modeling.

Explosive effects calculations are not precise; they depend on the type and nature of the bomb, on the orientation of the structure, reflective surfaces, etc. Modeling relies on assumptions and idealized environments. A primary assumption is the amount and type of explosive that will be used, where and at what orientation to the structure it will be positioned. As stated above, all assumptions in relation to bombs are questionable. It is not possible to predict exactly where the perpetrator will place the bomb and a variation of a few meters in relation to façades and reflecting surfaces can noticeably alter the explosive effects.

There is a likelihood scale that reflects a range of probabilities which can be addressed through Probabilistic Risk Assessment[98] which can assist in determining which possible attack vectors should be modeled.

Blast effects are dependent upon numerous variables many of which may be 'assumed out' in modeling. In some cases only the pressure and impulse effects are modeled and reported

---

95   US Army TM 5-855-1

96   A search of "Validation of CONWEP" will produce many studies validating the model against large and small scale tests.

97   See US Department of Veterans (2007) Paragraph 7.5.2 and FEMA 428 Chapter 4

98   Stewart (2012)

without recognition of the thermal or fragmentation effects and their impact on the structure. The angle of incidence is often assumed to be normal i.e. at 90 degrees and the façade assumed to be a linear plane, sometimes of infinite proportions. Fixtures, fittings and architectural design elements which create spaces between structural elements or reflective surfaces are sometimes assumed out of the calculations. Such assumptions may be suitable for generic modeling and appropriate in many cases. Specific detail on how a particular structural element will respond may require greater clarity on the assumptions made. Requests for advice which require an explanation of the variables and assumptions used (and how they are considered in the modeling) will be of value in helping determine the best responses.

Modeling often does not include damage from fragmentation which is difficult to predict in relation to bombs. Fragmentation calculation requires assumptions on the nature of the casing/vehicle, the age and condition of the casing/vehicle, and the orientation of the casing/vehicle. For a VBIED it is whether vehicle windows are open, the placement of the device in the vehicle, the nature of the explosive used, the method and direction of detonation and the ballistic stability of every fragment in relation to its ejected trajectory[99]. Also to be considered are nearby fittings, street furniture, and other objects which may become secondary projectiles. None of these can be predicted with accuracy.

What can be stated is that fragments will travel at high speed in an approximation of a straight line until deflected by an item or brought to ground by gravity. Fragments can be expected to travel further than the blast effects and to subject people and structures to impact damage.

Single degree of freedom calculations or other instances where only selected elements of the explosive effects are modeled assist in understanding the detailed response of specified components but do not reflect the multiple reactions and interactions occurring simultaneously on the real structure.

Of assistance in helping understand the complexities of explosive effects are images resulting from the modeling showing how explosive effects vary depending on the angle of incidence, distance and location along the structure.

For guidance on selecting consultants to assist with defining and scoping blast modeling see Chapter 10.

Charge Weight/NEQ

The specifics of which charge weights are to be used as the basis for calculations that are issued by governments tend to be classified. US Unified Facilities Criteria DoD Minimum Antiterrorism Standoff Distances for Buildings UFC 4-010-02 is restricted to

---

99   For example: Morrison and Williams 2003

those approved by the US Government. UFC 4-010-02 is referenced by other US DoD and FEMA documents[100].

The weight of the explosive excluding packaging and fragmentation casing is referred to as net explosive quantity (NEQ)[101]. The NEQ is usually expressed in terms of x kg of TNT equivalent.

Common NEQ provided by government agencies are: 23kg/50lb., 225kg/500lb., 500 kg, and 5000 kg. The historical, intelligence-led basis for nominating these as standard or common explosive charge weights has not been identified.

A review of open source media reports on bombing incidents around the world as well as accessible official data suggests that most bombs are less than 5 kg, most vehicle-borne IEDs (VBIEDs) have a NEQ of less than 20 kg, a few in the hundreds of kg and very few in the tonnes[102]. The normally nominated charge weights of 23, 225 and 500 kg are at the higher end of what experience would indicate is probable.

One method of considering the physical dimensions of NEQ for potential attack vectors and delivery methods is to visualize that a 5 kg weight can be held with an outstretched arm, 10 kg can be carried by the side of the body, 20 kg is a heavy two arm carry and anything above that will be transported on wheels.

On sites where there is a free flow of people and goods, consideration should be given to the maximum size of an explosive charge that can reasonably be brought on-site (hand carried, on wheels or by vehicle).

Once the NEQ is given as being above (say) 100 kg, and depending on the nature of the explosive used, there may be issues with the explosive being properly primed to ensure complete detonation of the material. This is particularly so when low sensitivity explosives or homemade explosives are used.

If modeling is requested for NEQ of hundreds of kg inside the building it should be tempered with consideration that bringing that quantity of explosive into the site should not be possible, given effective security capabilities.

Underground car parks and internal loading docks are exceptions and create a special set of security considerations. Under-building car parks permit the free flow of vehicles capable of carrying large quantities of explosives into the site. The construction of carparks creates numerous reflective surfaces which will augment blast effects. Building utilities and critical services often run through carparks and are exposed to damage from

---

100 As well as the United States Department of Defense documents considerable guidance is provided in Federal Emergency Management Agency (FEMA) and UK Centre for the Protection of National Infrastructure (CPNI) references.

101 See NATO AAP-6 (2007)

102 Author's collection of daily news articles relating to bomb incidents over 15 years

accidental or deliberate events.

Nominating explosive charge weights that may not be relevant to the environment or the probable attack vectors will result in recommended protective measures that are probably excessive and expensive. Defining realistic NEQ is fundamental to appropriate protective structure advice. Consideration should be given to what is 'probable' rather than 'possible'.

Unclassified evacuation distances for bombs are published by government agencies such as the US National Counterterrorism Centre, the Bureau of Alcohol, Tobacco, Firearms and Explosives, the National Ground Intelligence Centre and the Australian Bomb Data Centre. These evacuation distances are provided as indicators of safe distances from blast and fragmentation effects not as blast/distance effects tables and therefore are not relevant to blast analysis for structural response.

Type of Explosive Modeled

Another factor in modeling is the use of TNT equivalency. While recognised as the standard for explosive effects, TNT does not reflect the reality of bomb construction. The velocity of detonation, brisance, impulse and temperature differ with each type of explosive. TNT is difficult to obtain as it is rarely used in the commercial and military sectors other than as a component of other explosives.

Calculations that utilise the type of explosive most likely to be employed by perpetrators may be of greater value. Consideration of the size of the expected bomb would assist in selecting an appropriate explosive to model. For small devices Pentolite[103], TATP[104], or low velocity propellants may be appropriate. For larger devices ammonium nitrate/fuel oil (ANFO) or other nitrate-based explosives (commercial or home-made) may be more relevant. One hundred kg of ANFO, which is a more probable bulk explosive than 100 kg of TNT, will have different effects on the structure.

Consideration of non-industrial (home-made explosives HME) makes modeling and testing difficult, as it is not possible to forecast the quality or consistency of HME. This is reflected in the reliance on well-documented and validated explosives and their effects such as TNT regardless of the relevance to the probable attack vectors.

Table 4 shows the difference in peak reflected pressures (Pr) for TNT, Pentolite and ANFO for various NEQ and ranges.

103 Pentolite is an example of a reasonably common commercial high explosive.

104 See the Glossary

Kg	Range m	ANFO Pr	TNT Pr	Pentolite Pr
5	5	297	354	421
10	5	561	679	819
100	20	119	137	158
225	20	221	260	307
1000	20	867	1056	1280

Table 4 – Comparison of explosives
(Pr = Peak reflected pressure)

Use of Results

The results of blast modeling can be used to:

- determine if critical elements of the site are at risk should a bomb of x size be placed at y distance
- identify additional physical controls which may be implemented to limit what can be placed near the site or critical elements,
- identify additional procedural controls which may be used to limit what can be placed near the site or critical elements, and
- identify whether physical strengthening of the structure could limit the consequences of an explosion.

Summary

Modeling of blast effects is of great value in determining the possible effects of a bomb against a structure.

It must be noted that all assumptions in relation to bombs are questionable.

The reason for the modeling must be clearly understood and explained to those conducting the modeling.

Depending on the required information either Detailed Computer modeling or Conventional modeling may selected.

Poorly defined requests for modeling, a lack of understanding of the limitations of modeling and inappropriate modeling will result in flawed results.

The underlying concept should be the modeling of realistic bomb attack vectors given the existing physical and procedural controls.

Annexes

Annex A

Summary of Key Concepts and Principles

The key concepts and principles presented in this book are:

Bombs are devastating.

Bombs are used by criminals and terrorists for personal and political gain.

The four types of bomb incidents, in order of likelihood are:

- Unattended items
- Threats
- Bombs (including Hazardous mail)
- Post-blast

Organizations can assess and allocate resources for bomb incident management by using risk management principles.

Managers can reduce the risks posed by bomb incidents by implementing appropriate policies, procedures, training and where appropriate, technology.

The fundamental principle of bomb security is the ability to keep the bomb some distance from the asset and the ability to move the assets, particularly people, away from the bomb.

Organizations can implement procedures to

- Assess and respond to threats and unattended items,
- Identify and respond to bombs including safe and appropriate evacuation,
- Respond to a post-blast situation.

All staff need to be aware that a bomb incident may occur and they need to know how to **Receive, Record,** and **Report.** Selected managers need to be trained to **Review** and **Respond.** This is the "5 R" mnemonic.

Selected volunteer staff need to be trained in basic search techniques to assist with gathering information to assist with threat assessments.

Some staff need to be trained how to assess unattended items to determine if they pose a hazard or are rubbish or lost property. Use of the **VALID** methodology will assist in assessing unattended items to identify which may pose a hazard while minimising disruption.

It is possible to provide additional protection from bomb incidents by implementing appropriate policies, procedures and training.

Where applicable, it may be possible to provide physical protection by limiting what can be brought into close proximity with critical elements of the site and by hardening specific structural elements.

Modeling of potential bomb incidents is of value as long as the information being sought is clearly defined and the limitations of modeling are recognised and accepted.

Bomb incident management should be integrated and coordinated with other management plans relating to emergencies, business continuity/resilience, work place safety, human resources, building/facility management, environmental management, media management, etc.

The aim of bomb incident management is to protect life while minimising disruption.

Annex B

Example Index of a Bomb Incident Management Plan

Introduction

Distribution & Version Control

Aim/Intent

Manager/Section Responsible for the Plan

Related Organizational Plans/Procedures

Principles of Bomb Incident Security

Types of Bomb Incidents

Bomb Threats

    Background/Context

    Receiving and Recording the Threat

    Evaluating the Threat

    Time Available to Evaluate the Threat

    Threat Response Options

Unattended Items

    Background/Context

    Identifying and Reporting

    Evaluation of Unattended Items (possibly including VALID methodology)

    Unattended Item Response Options

Bombs (IEDs)

    Identifying a Bomb

    Immediate Response

Mail Bombs (and Other Hazardous Mail)

    Hazardous Mail Identification

    Hazardous Mail Investigation

    Hazardous Mail Response Options

Searches

    Purpose

    Organization of Search Teams

    Command/Control

    Training

Evacuation Considerations (Bomb Incident)

Evacuation

Safety

Security

Post-Evacuation and Re-occupation

Alignment with Emergency and other Procedures

Incident Recording and Reporting Procedures

Post-Blast Considerations

Short Term

Medium Term

Long Term Training (who is responsible for delivery, frequency, recording, validation)

Testing/Exercises (types, frequency, relation to Fire and other emergency responses)

Checklists

Receiving, Recording and Reporting of Threats – All staff

Receiving, Recording and Reporting Unattended items – All staff

Review and Response responsibilities and options

Search team organization, communication and techniques

Post-blast – immediate and medium term actions

Contact Lists Internal and External to the organization

Annex C

Examples of Bomb Incidents

The following are based on real-life incidents and are provided as examples of bomb incident management.

1 Unattended Item

At an international airport an unattended bag was found on the secure side (air side) of the passenger screening point. The bag was a shopping bag with the name of one of the clothing and accessory retailers located in the airside concourse. The security supervisor approached the bag, looked inside and identified a brand new red handbag. He picked up the bag and took it to the retailer to see if they could help identify who purchased it so the owner could be paged or tracked to a flight.

Comment: Traditional wisdom might suggest that the supervisor should not have touched the bag and should have initiated an evacuation until the owner of the item could be identified. In the incident described, the supervisor made the sound assessment that the item was most probably an item of shopping left behind by a traveler or visitor. There were no indicators to suggest the item posed a hazard. Initiating even a partial evacuation would have disrupted the operations of the airport and airlines. If the supervisor was concerned about moving the item before it had been investigated, alternatives might have included posting a guard near the item while the CCTV footage was reviewed or a staff member from the shop be asked to attend and identify the handbag as a stock item they sold.

2 Unattended Item

At an international airport an unattended briefcase was found in a public area. A call for the owner was made over the public address system. Concurrently an investigative team arrived and X-rayed the item and identified batteries and other items, they declared the item 'suspicious' and requested the bomb squad attend. A complete evacuation of the area was initiated. During the evacuation a man approached airport staff, claimed ownership of the item and was able to describe it and its contents (which were not hazardous). The man was refused access to the area until the item had been 'rendered safe' by the bomb squad. The period of disruption to the airport and airlines was hours.

Comment: The risk of the item being hazardous should have been reassessed when the owner made himself known. Once the owner was identified he should have been allowed to claim the item. If the investigating team was concerned they could have checked his details against any name tags or other marking on the briefcase or questioned him further on the contents. The response process had gained an impetus that was hard to stop.

3 Bomb Threat

A manager acting in a senior executive position for a major company received a bomb threat on a private (unlisted) office number. The threat contained specific in-house information and the threat was credible. The company had effective, documented and practiced bomb threat procedures. The manager was not aware of the company procedures and proceeded to apply management analysis techniques to the problem, this resulted in a 15 minute delay before security was informed of the threat and the company's threat assessment procedures were implemented.

Comment: The ability to respond effectively to the threat was reduced because the manager was not aware of the procedures. There was a company bomb threat card in his desk but he either had not been trained in its existence and use or had ignored the training. The delay may have endangered the lives of staff and visitors and the organization's operations.

4 Unattended Item

At a hotel, during a major international event, a guest heard a strange beeping noise coming from a rubbish bin. He informed a policeman who approached the bin and recognised the distinctive sound of a low battery alert from a smoke detector (as used in that country). He searched through the bin found the smoke detector and removed the battery. It is surmised that one of the guests being annoyed by the sound removed the smoke detector from his room and put it in the rubbish.

Comment: The knowledge and experience of the policeman allowed him to respond appropriately to this incident by identifying the item as non-hazardous. Someone without the same experience may have decided on a different course of action. If the policeman had found a bomb or other item in the rubbish bin then he would have recognised it as hazardous and initiated an evacuation.

5 Unattended Item

At a major public event a member of the public raised the lid of a rubbish bin and saw a brand new sports bag. She identified it as being unusual and not fitting the environment. She notified a policeman who immediately initiated an evacuation away from the bin. A bomb squad, which was on standby at the event, investigated and found the bag was torn and empty. It is surmised that a member of the public had discarded the torn bag.

Comment: In this case evacuation was appropriate as a new bag is not expected in the rubbish. The likelihood of identifying a potentially hazardous item was increased due to an aware member of the public. The prompt response was possible because planning had foreseen that the event was of a nature that could be vulnerable to bomb incidents and the consequences were assessed as being high enough that a bomb squad was on location to reduce the likelihood of an identified bomb detonating.

6 Bomb Threat

Over 1000 Iraqis were killed in one incident in September 2005 while on a pilgrimage to the Kadhimiya mosque. They died in a crowd crush which was caused by a rumour of a suicide bomber in the crowd. This is a larger number than killed by any bomb.

Comment: The rumour was believed because many thought it possible that a suicide bomber could have joined the procession. How the likelihood of a bomber joining the crowd or how the consequences could have been mitigated if the bomber was identified makes an interesting and challenging risk analysis exercise.

7 Mail Bomb

During a mail bomb campaign, a senior staff member in the targeted industry received a package which was incorrectly addressed and which he did not recognise. As he left the office he jokingly said to his secretary "I will open this outside in case it is a bomb". He was subsequently killed while opening the parcel when the device proved to be a bomb.

Comment: It appears the senior staff member was aware that his industry was the target of a mail bomb campaign. While it is possible he identified the item as potentially hazardous, at least sub-consciously, the risk of 'failing to respond appropriately to a mail bomb ....' was realised with dire consequences. Once an item is believed to be hazardous, staff must know how to implement the appropriate response measures, even if it is only to notify the Emergency Manager.

8 Post Blast

An embassy was the target for a bomb, which, because of the security risk assessment and subsequent security measures, exploded outside the fence-line. Although the embassy had an effective 'hold in place' procedure the staff were immediately evacuated outside in accordance with the fire drill procedures.

Comment: All the staff inside the embassy survived the initial blast. By sending them outside they were exposed to hazards resulting from the initial explosion and possibly from other bombs. Although appropriate response procedures for external incidents had been developed it appears that staff, in particular the Emergency Control Organization, were not aware of these procedures and the consequence mitigation treatments that had been put in place were ignored.

9 Bomb Threat

During the build-up for the closing ceremony for the 2000 Sydney Olympic Games, a threat was received stating that explosives had been placed on the Harbour Bridge and that the closing ceremony should be cancelled. The threat (one of many during the Games) was assessed in accordance with the processes that had been developed. In addition to the heightened security, there were considerable quantities of pyrotechnics mounted to the bridge and all explosives, firing systems and other areas had been carefully checked and rechecked by a number of contractors and agencies.

Comment: Because of the physical and procedural risk mitigation treatments in place the assessors were able to determine that the likelihood of the perpetrator having done what they claimed was extremely low. As a result there was no requirement to disrupt the event. If required the threat assessment team had additional risk assessment measures available including: reviewing access control systems, reviewing CCTV and sending out search teams trained for that particular environment.

10 Bomb

The bombing of the United States military accommodation building at Kohbar towers in Saudi Arabia on the evening of 25 June 1998 was limited in its effect because of the risk management measures put in place. The bomb is estimated to have contained approximately 10 tonnes of explosives. The size of the bomb was a response to the access control measures that denied the perpetrators the opportunity to get close to the building. There were guards on patrol with visibility of the perimeter, one noticed the tanker truck and determined that it was out of place and potentially posed a hazard. An evacuation was immediately initiated. Only 19 service personnel died.

Comment: Without the ability to detect and respond to the bomb the casualties would have been considerably higher, possibly greater than the bombing of the Beirut Marine barracks in 1983 when 241 service personnel died.

11 Bomb Threat

A government storage and logistics facility received a bomb threat at lunchtime every Friday. The management response was to allow all staff to leave the site with their vehicles and close the site until the following Monday. The site lost a half day's productivity every week. On one Friday, after the threat was received, all staff were evacuated and assembled in the car park, which had been searched. Specialists then searched the entire site, this took until late evening, at which time it was determined that it was safe for the staff to leave. No further bomb threats were received.

Comment: No threat evaluation was conducted. The impact to the organization was disruption to operations due to poor bomb threat response procedures. A bomb threat process was introduced and a method of moving the consequences of the threats to the staff (and probably the perpetrator) was identified.

12 Bomb Threat

A university always received bomb threats when exams were being held. Initially the response was to evacuate the building under threat. The consequences included significant disruption to students and staff, having to reschedule exams, having to rewrite or have prepared alternative exams, and alterations to students travel plans including for overseas students. A bomb threat assessment process was introduced; the exam rooms and surrounds were searched prior to the exams; bags and other items were not permitted in or near the exam rooms making searches by security staff in response to threats quick and simple.

Comment: As a result of the ability to assess the risk posed by a threat, it was determined that it was unlikely that a perpetrator could have done that which they claimed. Thereafter it was rare that an evacuation was called and the threats stopped as the desired results (evacuation and disruption) were not forthcoming.

13 Bomb

A significant IT facility, the loss of which would have had national consequences, had extensive access controls into the mainframe rooms. The external wall of the mainframe room opened onto the visitors' car park over which there were no controls. It was assessed that the computers were vulnerable to a bomb being placed in close proximity to the wall. The likelihood was reduced by restricting access to that part of the car park.

Comment: As with many security risks, a risk may require a number of treatments and a treatment may help mitigate a number of risks. In this case altering the car park reduced the likelihood of damage to the asset from a range of hazards including a vehicle driving into the wall.

14 Bomb

During a major international event with large public crowds it was proposed that all rubbish bins should be removed to prevent bombs being hidden in them.

Comment: A proper risk review was conducted and the risk of a bomb being concealed in a bin was assessed for its likelihood and consequences given that operating environment. The risk was then expanded to include all locations where a bomb could be concealed including street furniture and gardens and whether these should also be removed. As there was already a high level of overt and covert observation of the area, frequent searches of the site and pre-planned response options were included in the threat evaluation. A risk analysis was also conducted into the consequences of removing the bins; risks relating to infection, infestation, trip hazards, damage to the environmental image of the event and litigation were identified. As a result it was decided that removing the bins increased other risks to unacceptable levels. Instead of removing the bins, they were emptied more often, the visible security presence in the area was increased, and alternate styles of bins were used in 'high risk' areas. It should be noted that by decreasing one risk other risks may be generated or altered.

15 Bomb - Training Device

A major sporting stadium evacuated 75,000 people when a bomb was found in a men's bathroom. The bomb was later identified as a training device that had been used during an exercise to train search dogs four days before. Media images of the device showed a pipe with a mobile phone and wires taped to it.

Comment: A number of issues are raised by this incident. Firstly there is no requirement for training devices to look like bombs; people should be trained to look for that which is out of place. Items used for training must be recorded when placed and recovered to ensure none are left behind. Media reports state that the device did not have any explosive 'scent' which is why it was not detected and recovered, this raises questions over why it was used for the training. If the intent was to help the dog handlers identify items not detected by the dogs then any item that did not fit the environment would have sufficed. Media reports suggest the cost to the hosting club would exceed three million British pounds (~US$4.4 million).

Annex D

Glossary

The following definitions are used in this book; alternative definitions may be used by national agencies.

TERM	DEFINITION
Ambush Device	Devices placed along exit routes, assembly or other areas with the intent of killing those leaving the site or passing the bomb. These attacks bring the victims to the other bomb(s).
ANFO	Ammonium Nitrate Fuel Oil explosive with a TNT equivalency of ~0.8. A common base for many mining and agricultural explosives. Ammonium Nitrate by itself can be detonated with a TNT equivalency of ~0.3.
BCP	Business Continuity Plan sometimes also referred to or aligned with the Business's Resilience Plan.
BDC	Bomb Data Centre
BoH	Back of House, logistic, administrative, operational, utility and plant areas not usually accessible by the public.
Bomb	A generic term for IED. Note: "Bomb" may also be used to describe an item of military ordnance such as an aerial or mortar bomb.
Brisance	The shattering capacity of explosives, related to the detonation pressure. Lower brisance explosives are used to 'push' e.g. in mining operations, high brisance explosives are used to 'shatter' e.g. in artillery shells.
Booby Trap	See Victim Operated
CFD	Computational fluid dynamics – computer-based assessment to determine effects against defined finite elements of a structure at given times from the start of an event. Used for detailed blast effect analysis.
Charge Weight	See 'NEQ'.

TERM	DEFINITION
Cloak Room	A room for temporary storage of coats, bags, umbrellas, etc.
CONWEP	Conventional Weapon Effects software as presented in US Army Technical Manual 5-855-1. Used for blast effect analysis.
Dirty Bomb	A bomb that contains radiological material and is designed to contaminate an area and any victims. Some authors may use the term to include IEDs containing chemical or biological material.
ECO	Emergency Control Organization.
Emergency Manager	Person responsible for managing the response to an emergency. May also be referred to as Chief Warden Incident Controller, etc.
Emergency Services	Uniformed services, usually government operated, including law enforcement, fire fighting and emergency medical services
EMP	Emergency Management Plan
ERP	Emergency Rendezvous Point
FoH	Front of House, areas where the public or visitors attend and where activities related to the public and visitors occur.
Fuze	Also" "Fuse', the system used to initiate the detonation of a bomb
HAZMAT	Hazardous Material
Hazardous Mail	Any item sent through the mail system containing a potentially dangerous material, may be sent with or without malicious intent. In this book also taken to include courier-delivered items.
HME	Homemade explosive.

TERM	DEFINITION
IED	Improvised Explosive Device. US Department of Homeland Security definition: "a "homemade" bomb and/or destructive device to destroy, incapacitate, harass, or distract. IEDs are used by criminals, vandals, terrorists, suicide bombers, and insurgents. Because they are improvised, IEDs can come in many forms, ranging from a small pipe bomb to a sophisticated device capable of causing massive damage and loss of life. IEDs can be carried or delivered in a vehicle; carried, placed, or thrown by a person; delivered in a package; or concealed on the roadside."
kPa	Kilopascal. A measure of pressure. 1 kilopascal (kPa) = 0.145 pounds per square inch (psi) = 1 kilo Newton per meter squared (1kN/m2)
Leg Wires	The wires leading from an electric detonator to a firing system.
Marshal	See Warden.
Multiple Devices	Where the perpetrator places more than one bomb at the site. The intent being to cause maximum damage through multiple simultaneous, or near simultaneous explosions.
NEQ	Net Explosive Quantity, the weight of the explosive material without consideration of the casing or packaging, sometimes also referred to as Charge Weight.
Off Route Device	A bomb placed to the side of a traffic route and detonated to attack a particular target.
Person Borne IED	Person Borne IED.
Placed IED	A bomb brought into or near the site and placed by the perpetrator who then departs. May be identified as an Unattended Item.
Primary Fragmentation	Fragmentation from the casing of the bomb- may be augmented by the inclusion of nails, bolts or other metal items.

TERM	DEFINITION
Projected IED	A bomb that is fired at some distance towards the target, may be hand thrown or fired from a mortar of grenade launcher.
Pentolite	A composite explosive of PETN and TNT used in commercial and military sectors. Depending on the mixture has a VoD of ~7400 to ~7800 ms and a TNT equivalency of 1.33.
PETN	Pentaerythritol tetranitrate a common commercial explosive. Has a VoD of ~8400 ms and a TNT equivalency of ~1.6.
PRE Factor	Relative effectiveness factor, as a measure of comparison against TNT, see TNT equivalency.
Remote Controlled Device	A bomb that detonates on receiving a signal from a remote location. Remote triggering methods include: fixed wire, radio, phone, visible or infrared light,
RDD	Radiological Dispersal device, see Dirty Bomb
Secondary Devices	Placed to attack the responding emergency services. Usually deployed by groups that have decided that the emergency services, including bomb squads, are defeating their efforts and should be targeted. Placement of secondary devices usually requires the perpetrators to predict how the responding emergency services will deploy and where.
Secondary projectiles or fragmentation	Fragmentation caused by the breaking up and projection of items near the bomb but not from the casing of the bomb.
Secondary hazards	Those materials on-site that are safe until acted upon by an explosion.
Suicide device	A bomb delivered by a person who intends to detonate it while they are still in the danger area.
TATP	Triacetone triperoxide may be manufactured as a home-made explosive. Has a VoD of ~ 5300 ms and a TNT equivalency of ~0.83.

TERM	DEFINITION
Timed device	A bomb that is detonated by some form of timer after a pre-determined period has elapsed.
TET	Threat Evaluation Team. A team of personnel trained in threat evaluation processes.
TNT	Trinitrotoluene a high explosive. Has a VoD of ~6900 ms. Is the explosive against which others are measured and has a RE factor of 1.
TNT Equivalency	The measure of force applied by an explosive compared to TNT which is given a rating of "1" e.g. AFNO has a TNT equivalency of ~0.8.
VBIED	Vehicle Borne IED
Victim Operated	A bomb designed to function if touched, lifted, pushed or otherwise acted upon by a person.
VoD	Velocity of detonation, the speed at which the transition from a solid state to a gas occurs within an explosive, usually measured in 1000s of meters per second.
Warden	A staff member trained in and responsible for the safe movement of people during an emergency. Part of the Emergency Control Organization.

Annex E

Bibliography and Additional Reading

The following are referred to in the book or are recommended as additional reading.

Bloom M. (2005) *Dying to Kill: The Allure of Suicide Terror*, Columbia University Press.

Campbell C. (2002) *Fenian Fire, The British Government Plot to Assassinate Queen Victoria*, Hammersmith UK, HarperCollins.

Gebbeken N. F. (2015 December) *The role of civil engineers as first responders in disaster management* International Journal of Protective Structures Vol 2 No 4

Germershausen R. (Ed) (1982) *Book on Weaponry*, Düsseldorf, Rheinmetall GmbH.

Gips M. (1999 June) *Defusing Threats* Security Management Volume 43 No 6 p22.

Giusti C O'Hara S (1998 November) *Mail Centre Security* Security Management Volume 42 No 11 p61.

Gould K. E, Tempo K, *High Explosive Field Tests – Explosion Phenomena and Environmental Impacts* US Defence Nuclear Agency, October 1981 Washington DC

Harowitz S. (Ed) (1997 February) *What's Your Mail Policy* - Survey Security Management Volume 41 No 2 p16.

Harowitz S. (1997 May) *Dangerous Deliveries* Security Management Volume 41 No 5 p38.

Hoad P. (2005) B*last Engineering – Preliminary Views*, Security Oz #34 Mar/Apr 2005, Australia Media Group.

Jenkins B. (1985) *Terrorism and Personal Protection*, Butterworth Publishers Boston, London.

Kitteringham G. (2007 September) *Down and Out in Record Time* Security Management Volume 51 No 9 p78.

L'Abbe R. et al Blast Effects Parts 1-4 (1989) The Detonator Volume 16 no 6 and Volume 17 nos, 2, 4 and 6. International Association of Bomb Technicians and Investigators Colorado Springs.

Lee J. (2007 January) *Gimme Shelter* Security Management ASIS International Volume 51 No 1 p52.

Mendis P. Crawford J. Lan S. (2005) *An Introduction to Explosion Effects and Design for Blast*, Sydney Australia, Advanced Protective Technologies for Engineering Structures (www.aptes.com.au).

Narayanan T. V. (1996) *Modern Techniques of Bomb Disposal and Detection* RA Security

System, New Delhi

Netherton M. D. (2012) *Probabilistic Modeling of Structural and Safety Hazard Risks for Monolithic Glazing Subject to Explosive Blast Loads* University of Newcastle Australia

Newman G. (2005) *Bomb Threats in Schools* Community Orientated Policing Services US Department of Justice www.cops.usdoj.gov.

Ngo T. Mendis P, Gupta A, Ramsay J. *Blast Loading and Blast effects on Structures – An overview* Electronic Journal of Structural Engineering – Special Issue Loading on Structures 2007.

O'Toole M. (2000) *The School Shooter: A Threat Assessment Perspective Critical Incident Response Group* Federal Bureau of Investigations Academy Quantico http://www.fbi.gov/publications/school/school2.pdf.

Pape R. (2005) *Dying to Win: The Strategic Logic of Suicide Terrorism*, Subscribe Publications.

Read I. (1942) *Explosives*, UK, Pelican Books Limited.

Remennikov A. (2003) *A Review of methods for Predicting Bomb Blast effects on Buildings*, University of Wollongong

Richardson L. (2006) *What Terrorists Want, Understanding the Terrorist Threat*, John Murray Publishers.

Russo G. (2001) *The Outfit, The Role of Chicago's Underworld in the Shaping of Modern America*, Bloomsbury

Seuter E. (1997 May) *It's in the Mail* Security Management Volume 41 No 5 p28.

Seuter E. (2003 July) *The Dog Days of Detection* Security Management Volume 47 No 7 p49.

Sharp A. (2005), *From Dublin Castle to Scotland Yard: Robert Anderson and the Secret Irish Department*, http://whitechapelsociety.com/Journal/oct05 .

Smith P. (1995) *Blast: Fundamentals of Blast Loading and Structural Response Analysis*, UK, Royal Military College of Science, Cranfield University.

Stewart M. G. (2008) *Cost Effectiveness of Risk Mitigation Strategies for Protection of Buildings against Terrorist Attack* Journal of Performance of Constructed Facilities March/April 2008

Stewart M. G, Netherton M. D, Shi Y and Grant M *Probabilistic terrorism risk assessment and risk acceptability for infrastructure protection* Journal Of Performance of Constructed Facilities Vol 3 No 1 2012

Sullivan J. Bunker R. Lorelli E. Seguine H. Bergert M. (2002) *Unconventional Weapons Response Book* First Edition, Jane's Information Group.

Talbot J. Jakeman M. (Ed) (2007) *Security Risk Management Body of Knowledge* Canberra Australia, Risk Management Institute of Australia.

Williams C. (2004), *Terrorism Explained*, New Holland Publishers.

Williams D. (2004), *Big Bang Theory - Dealing with Bomb Threats*, Australian Leisure Management February/March 2004 Sydney, Quality Images Publishing.

Yates, A. (2012), *Blast: How explosive devices kill people and destroy buildings*, Australian Security Research Centre.

Government Publications

Australian Emergency Management Authority (2004) *Emergency Management for Public Venues in Australia – Guidelines*, Emergency Management Australia, Canberra.

Australian Federal Police, Australian Bomb Data Centre (2009) *Bombs Defusing the Threat*, Australian Government, Canberra.

Australian Federal Police, Australian Bomb Data Centre (1995) *Evaluation of Shatter Resistant Films*, Williams D. (ed) Australian Government, Canberra.

Australian Federal Police, Australian Bomb Data Centre (1994) *Evaluation of Mail Screening Equipment*, Williams D. (ed) Australian Government, Canberra.

Australian Standard 3745 *Emergency Control Organisations*.

Australian Standard Book 167 *Security Risk Management*.

Australian Standard Book 328 *Mailroom Security*.

International Standards Organization (ISO) 31000 *Risk Management – Principles and Guidelines*.

International Standards Organisation (ISO) 16933 *Glass in building -- Explosion-resistant security glazing -- Test and classification for arena air-blast loading*.

International Standards Organisation (ISO) 16934 *Glass in building -- Explosion-resistant security glazing -- Test and classification by shock-tube loading*.

New York City Police Engineering Security *Protective Design for High Risk Buildings*, 2009.

NATO Glossary of Terms and Definitions AAP-6 (2007) 2-1-2 page No. 130, Brussels http://www.nato.int/docu/stanag/aap006/aap6.htm.

Royal Mail Annual Report and Accounts 2006-07 ftp://ftp.royalmail.com/Downloads/public/ctf/rmg/RandA_2006-07_26-10-07_FINAL_ revised.pdf

TM9-1300-206, (1973) *Ammunition and Explosives Standards*, US Department of the Army.

UK Ministry of Defence (1954) Service Text Book of Explosives UK Government.

United Kingdom Centre for Protection of National Infrastructure (CPNI) *Protecting Against Terrorism* Third Edition, 2010.

United Kingdom Centre for Protection of National Infrastructure (CPNI) *Screening mail and deliveries*, 2016 http://www.cpni.gov.uk/advice/Physical-security/Screening/Mail-and-deliveries

UN Economic and Social Council (ECOSOC) Committee of Experts on the Transport of Dangerous Goods (TDG) *Transport of Dangerous Goods – Recommendations of Experts on the Transport of Dangerous Goods* (The UN Orange Book) (2005), 14th revised edition, New York and Geneva.

US Army TM5-855, (1992) *Conventional Weapon Effects (CONWEP)*, US Army Waterways Experiment Station, US Department of the Army.

US Bureau of Alcohol, Tobacco and Firearms, (1999) *Vehicle Bomb Explosion Hazard and Safety Distance Tables*, US Department of the Treasury.

US Bureau of Alcohol, Tobacco and Firearms, (1987) *Bomb Threats and Physical Security Planning*, ATF P 7550.2 7/87 US Department of the Treasury.

US Department of Defence National Ground Intelligence Centre, *Improvised Explosive Device Safe Standoff Distances Cheat Sheet*.

US Department of Homeland Security FEMA-426 (2011) *Reference Manual to Mitigate Potential Terrorist Attacks Against Buildings*, October 2011 Edition 2.

US Department of Homeland Security FEMA-427 (2003) *Primer of Design of commercial Buildings to Mitigate Terrorist Attacks*.

US Department of Homeland Security FEMA-428 (2003) *Primer to design Safe Schools Projects in Case of Terrorist Attacks*.

US Department of Justice and Department of Homeland Security (2015) *Bomb Threat Guidance*.

US Government Accountability Office (2005) *US Postal Service Guidance on Suspicious Mail Needs Further Refinement* GAO-05-716 US Government.

US General Services Administration USA *Standard Test Method for Glazing and Window Systems Subject to Dynamic Overpressure Loadings*.

United States Unified Facilities Criteria Department of Defense *Structures to Resist the Effects of Accidental Explosions (UFC 3-340-01)*.

United States Unified Facilities Criteria Department of Defense *Structures to Resist the Effects of Accidental Explosions (UFC 3-340-02)* 5 Dec 08. Note: supersedes US Army TM5-1300

US Postal Service *Best Practice for Mail Centre Security*
http://www.usps.com/communications/news/security/bestpractices.htm.

US Postal Service *Postal Facts 2008*
http://www.usps.com/communications/newsroom/postalfacts.htm.

WorkCover (2002) *Guidelines for Assessing the Risk of Exposure to Biological Contaminants in the Workplace* Sydney Australia, New South Wales Government.

Conference Papers and Proceedings

Mendis P. Lai J. Dawson E. (Ed) (2006) *Recent Advances in Security Technology*, Proceedings of the 2006 RNSA Security Technology Conference, Canberra, Australia, Research Network for a Secure Australia, Canberra.

Morrison R. Williams D. (2003) *Effects of Large Vehicle Bombs*, International Association of Bomb Technicians and Investigators, Australian Chapter Training Conference, Canberra. Delivered by D. Williams on at the IABTI International Training Conference of 2004.

Stewart M. *Reliability and Risk reduction of Blast Measures for Infrastructure* Blast design and Modeling Forum November 2012, University of Newcastle Australia

Swizdak M. Tatom J. Tancreto J. (2005) *SCIPAN—A Program to Determine the Effects of Blast Loading on typical Structures—Update* Directorate of Explosive Safety Parari Conference Australian Department of Defence, Canberra.

Williams C. (2005), *Suicide Bombers presentation to 2005 Securing Australia Conference*, Australian Homeland Security Research Centre, Canberra.

Williams D. (2001), *Bomb Incidents - Reducing Likelihood and Consequences*, International Police Conference, Adelaide, Australia.

Williams D. (2003), *Using a Study of Motives to Determine the Type of Bomb that may be Used*, The Australian Police Summit 2003, Sydney, Australia.

Williams D. (2003), *Explosives Tool or Weapon*, Australian Ordnance Council Parari Conference Australian Department of Defence, Canberra.

Other Resources

Centre for Chemical Process Safety, *Guidelines for Evaluating the Characteristics of Vapour Cloud Explosions, Flash Fires and BLEVEs*, American Institute of Chemical Engineers, New York 1994

NIJ Guide 103-00 (2001) *Guide for the Selection of Chemical and Biological Decontamination*

Equipment for Emergency First Responders, Washington, US Department of Justice. http://www.mi5.gov.uk/output/Page37.html.

Notes

Notes

Notes

Notes

Notes

Notes

Notes

Notes